Telecommunications Primer

FOURTH EDITION

Graham Langley MBE, BSc, CEng, FIEE, MBIM

Fourth edition prepared by

John Ronayne BSc, CEng, FIEE

D1378441

PITMAN
PUBLISHING

Pitman Publishing
128 Long Acre, London WC2E 9AN

A Division of Pearson Professional Limited

© G. A. Langley & J. P. Ronayne 1993

First published in Great Britain 1983
Reprinted 1984
Reprinted 1985
Second edition 1986
Reprinted 1986, 1987, 1988
Third edition 1990
Reprinted 1991
Fourth edition 1993

ISBN 0 273 60157 1

10 9 8 7 6 5 4 3

Typeset by Mathematical Composition Setters Ltd, Salisbury, Wiltshire
Printed in England by Clays Ltd, St Ives plc

Contents

Preface

More and more people are beginning to believe that future national prosperity will depend tremendously on the extent to which efficient use is made of telecommunications and computers. The computing and telecommunications industries have indeed in recent years moved together and a new term has been coined to describe this movement: Information Technology.

A general knowledge of Information Technology is nowadays essential for anyone who wishes to be an effective top executive in any field; corporate efficiency often depends more on the understanding and wise use of IT than on any other single factor. It is no longer commercial sense to leave such matters entirely to the technicians and engineers directly concerned.

This book has therefore been prepared with three main objectives. First, to give new technical recruits to the profession a rapid introduction and background to the various specialist activities and modern developments which will be facing them, particularly in the telecommunications side of the IT field.

Secondly, to provide professionals in other disciplines, and non-technical managers generally, with easy-to-read but authoritative papers on many of the technically complex matters which are now becoming important parts of our everyday lives.

Thirdly, to give school-leavers and first jobbers an outline of what is going on in the IT world, so that they will not need to be treated like new-born babes when they step out into commercial life.

The book has been written in a user-friendly way which takes the reader step-by-step up the technological ladder; we have tried to ensure that no concept is mentioned until it has been defined in layman's language.

Introduction to the Fourth Edition

At the time of his death in October 1991, Graham had just completed the first draft of a new book, "The Business Telecommunications Primer". In discussing publication plans with Pitman, two things were obvious: Telecommunications Primer urgently needed a new edition and much material that could be used for it was included in the draft of Business Telecommunications Primer.

We decided therefore to concentrate on the new edition and to use Graham's business material to extend the book to embrace new subjects relevant to the business user. At the same time we have confined the subject area more closely to two-way communications and removed much material on broadcasting, particularly television, and basic electronics which is more adequately covered in other works. Television is still included but only in so far as it is relevant to two-way telecommunications since the television receiver is becoming one of the alternatives for the communications terminal bringing two-way communications to the home and the place of work.

Since the third edition, Graham in his draft and I, his new collaborator, have had to cover dramatic new developments in the telecommunications scene. These stem from two causes. There is the recent maturity in fibre optic technology making it possible to conceive of telecommunications bandwidths of hundreds of Megaherz. There is also the proliferation of mobile communications systems to the point where this too is maturing into a new and more extensive branch of telecommunications, namely personal communications, which is, and will increasingly, affect the whole of telecommunications.

Telecommunications is a field that seems to attract brilliant, interesting, nice people. Graham's text demonstrates that he is one of these.

It is a pleasure and an honour to be allowed to participate with Graham in this excellent book. It has been an objective to use Graham's new text and retain his existing text, together with his particular style wherever this was possible. The change between Graham and his collaborator is not seamless. Graham's writing is attractively personal. This has made it seem right for me to add my own views of what is good, bad and exciting in this wonderful technology.

Antibes, August 1992

Acknowledgements

In no way would I claim to be an expert in all the many subjects dealt with in this book, and there seemed little point in pretending to be re-inventing the wheel. A number of the sections given here do therefore make much use, with grateful acknowledgements, of material given in existing textbooks published by Pitman/Longman. In particular, several illustrations originally appearing in these books have been used in this book where applicable, and also a number of text passages used either directly or modified.

Although my debt to all these Pitman authors is considerable, I must record a special thank you to Mr P. H. Smale, whose *Telecommunications Systems* provided much basic information for many sections.

Another source of material, often consulted, was Telephony's *Dictionary* of telecommunications terms, published in Chicago in 1982. My grateful thanks go to the Telephony Publishing Corporation.

I should like also to thank colleagues in the British telecommunications industry who have helped by criticising first drafts and who provided germs for ideas used in these pages. Telecommunications these days is very much a team effort; suggestions have come from so many sources that to name all would be impracticable.

For the sections on satellite and maritime communications and for much good advice throughout I [JPR] have to thank Malcolm Leak, a colleague in Dar Al Handassah Consultants. Many other colleagues on Project Teams at ETSI have also been of great assistance. STC Submarine Systems provided valuable updating information on submarine cables. I am indebted to two other books for ideas for new illustrations, these are:

ISDN Explained, John M Griffiths, Wiley.

Racal Interlan on Interoperability, Richard Bowker, Racal Interlan.

Part A
Basic Principles

1 Telecommunications Systems

Telecommunications has been most neatly defined as the technology concerned with communicating at a distance.

The first requirement is for the original information energy (such as that of the human voice, or music, or a telegraph signal) to be converted into electrical form to produce an electronic information signal. This is achieved by a suitable transducer, which is a general term given to any device that converts energy from one form to another when required.

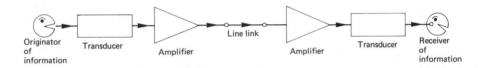

Fig. 1.1
Basic requirements for a one-way line telecommunication channel

In a line telecommunication system (fig. 1.1), the electronic signal is passed to the destination by a wire or cable link, with the energy travelling at a speed of up to 60% that of light (depending on the type of line). At the destination, a second transducer converts the electronic signal back into the original energy form. In practical systems, other items may also be required; for example, amplifiers may be needed at appropriate points in the system. Amplifiers do not change the signal from one

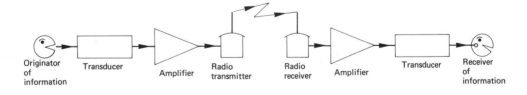

Fig. 1.2
Basic requirements for a one-way radio telecommunication channel

form of energy to another; they are usually inserted when it is necessary to increase the power level of signals to compensate for losses encountered.

For a radio system (fig. 1.2), a transmitter is required at the source to

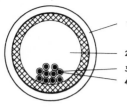

Unit with 10 fibres

1. Unit buffer
   tube
2. Filling
   compound
3. Coating
4. Fibre

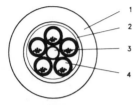

Unit—based cable with
50 fibres, formed from
one main unit containing
five 10—fibre units

1. Outer jacket
2. High—tensile
   armouring
3. Support member
   and cushioning
4. Unit with
   10 fibres

Fig. 1.3
Typical multi-fibre
cables as used in
distribution schemes

send the signal over the radio link, with the energy travelling at the speed of light, and a receiver is needed at the destination to recover the signal before applying it to the transducer.

At this point it is important to realize that, in both these systems, interference will be generated by electronic noise, and also that distortion of the electronic signal will occur for a number of reasons. These are undesirable effects and must be minimized in the system design.

Simple single-voice-band systems are one-way only (unidirectional), and generally called channels; domestic radio and television broadcasting are familiar examples of such systems.

Other systems, however, such as national telephone systems, must be capable of conveying information in both directions. To do this, the basic requirements must be duplicated in the opposite direction: a pair of complementary channels provide bi-directional communication, generally called a circuit.

When more than one circuit is needed between two points, it is not always economically practicable for another pair of wires or another radio system to be provided. Equipment is available which enables more than one voice channel to be carried on a pair of wires, a coaxial cable, a radio link or an optic fibre. Such multi-channel equipment is called multiplexing equipment.

Coaxial cable networks, with amplifying stations every few kilometres, now link together most of the cities in developed countries. These networks carry many thousand multiplexed channels.

Radio equipments operating at frequencies much higher than ordinary domestic radio sets and called microwave links are also able to carry thousands of multiplexed voice channels between terminals.

Optic fibres are a new and special form of transmission path in which energy representing many thousand voice channels can travel as pulses of light along a single glass or silica fibre comparable in diameter to a human hair (fig. 1.3). Optic fibre cable networks have now been installed in many countries, making possible a huge expansion of telecommunications services. Pulses of light travelling along fibres are not now used solely for short-distance telecommunications services as at one time seemed probable: current-generation transatlantic and transpacific submarine cables use optic fibres also; electric currents flowing along metallic conductors have suddenly become slightly old-fashioned.

2 Line System Characteristics

The simplest form of two-wire line is produced by using bare conductors suspended on insulators at the top of poles (see fig. 2.1). The wires must not be allowed to touch each other; this would provide a short-circuit and would interrupt communications.

Another type of two-wire line consists of conductors insulated from

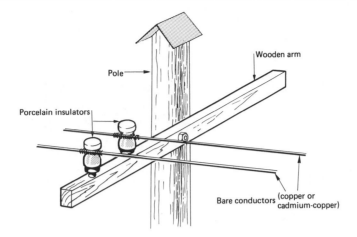

Fig. 2.1
Simple overhead
two-wire line

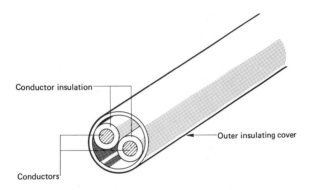

Fig. 2.2
Simple two-wire cable

each other in a cable, which also has an outer cover of insulation (see fig. 2.2). This outer sheath used to be made of lead but various types of plastic are now commonly used, particularly PVC. The two insulated conductors in the cable shown are often twisted together along the length of the cable, and are called a pair.

Many two-wire lines are often wanted between the same two places. These can most conveniently be provided by making a cable with a number of pairs of insulated wires inside it. Sometimes the wires are twisted together in pairs (as illustrated in fig. 2.3) but sometimes they are provided in fours or quads (as shown in fig. 2.4).

In order to identify the various wires, each wire has a colouring on the insulating material around it, in accordance with a standard colour code for cable pair identification.

As the frequency of an alternating current is increased, the current tends to flow along the outer skin of a conductor, and ordinary twin and quad type cables become inefficient. A special type of cable suitable for use at high frequencies has therefore been developed. This has one of its conductors completely surrounded by the second one, in the form of a tube. This type of cable is called a coaxial cable (shown in figs 2.5, 2.6, 2.7). The two conductors can be insulated from each other either by a solid insulant (or dielectric) along the whole length of the cable or by insulating spacers fitted at regular intervals as supports for the inner

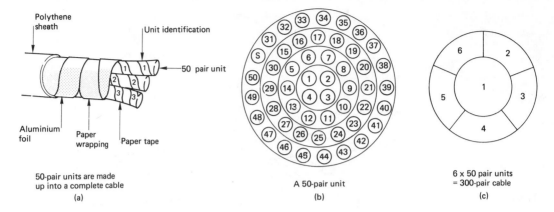

50-pair units are made
up into a complete cable
(a)

A 50-pair unit
(b)

6 x 50 pair units
= 300-pair cable
(c)

Fig. 2.3
Audio-frequency
unit-twin cable

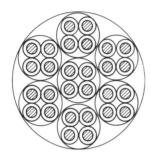

Fig. 2.4
Quad-type cable

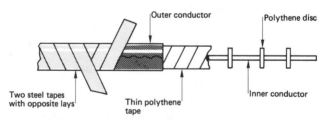

Fig. 2.5
Construction of coaxial
tube (air dielectric)

Fig. 2.6
Construction of flexible
coaxial cable (solid
dielectric)

conductor. In this case the main insulation is the air between the two conductors.

Whatever the type of cable used, the conductors always have some opposition to current flow. This is called resistance. Furthermore no insulating material is perfect, so the insulation used to separate the two conductors of a pair will always allow a very small current to flow between the two conductors, instead of all of it flowing along them to the distant end.

Also, the insulation between the conductors forms a capacitance which

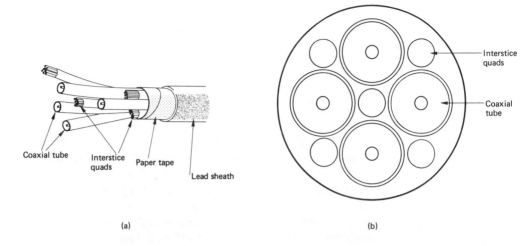

Interstice quads

Coaxial tube

Coaxial tube Interstice quads Paper tape Lead sheath

(a)

(b)

Fig. 2.7
Construction of
four-tube cable

provides a conducting path between the conductors for alternating currents (a.c.). This capacitance also has the ability to store electrical energy. The higher the a.c. frequency of the information signal, the more current travels across this capacitance path and the less reaches the distant end of the line.

When an electric current flows along a wire, a magnetic "field" is established around the wire. Bringing an ordinary magnetic compass near to a wire shows whether or not the wire is carrying a direct current; if it is, the magnet's needle swings. There is no need to cut the wire and insert a test meter. Whenever the current in a wire changes, either by the switching on-or-off of a one-way or direct current, or the repetitive changes of an alternating current, its accompanying magnetic field is made to change also, and energy is needed for these changes. This is called the inductance of the circuit. If there are other wires nearby, they are affected by such changing magnetic fields; there is inductive coupling between the wires. This is the principle used in transformers, which enable electric power to be transferred from one circuit to another without actual physical contact between the two circuits.

Energy is used up to make the current flow against the resistance along the conductors, and against the insulation resistance between the conductors. Energy is also used in charging and discharging the capacitance between the conductors. In multi-pair cables there is capacitive and inductive coupling between pairs also, so that some energy is passed from one pair to another. These losses further reduce the amount of energy that reaches the end of the original pair and so contribute to the total loss.

In the case of an information signal, all this lost energy has to come from the signal source, so the energy available gradually decreases as the signal travels along the line. This loss of energy along the line is called attenuation. If the line is long and the attenuation is large, the received energy may be too weak to operate the receiving transducer — unless some corrective action is taken.

3 Radio System Characteristics

When a radio-frequency current flows into a transmitting antenna (aerial), power is radiated in a number of directions in what is called an electro-magnetic wave. This is a complex signal with the same general characteristics as light but of a lower frequency; electro-magnetic radio waves travel at the same speed as light and can be reflected and refracted just as light can be. Some antennae are designed to be highly directional, some are omni-directional. The radiated energy (see fig. 3.1) will reach the receiving station by one or more of five different modes:

1) Surface wave   4) Via a satellite
2) Sky wave   5) Scatter
3) Space wave

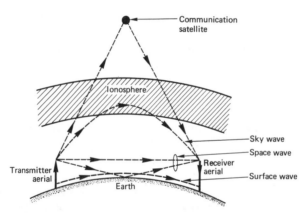

Fig. 3.1
Radio propagation
methods

The surface wave is supported at its lower edge by the surface of the earth and is able to follow the curvature of the earth as it travels.

The sky wave is directed upwards from the earth into the ionosphere (100 km or more above ground level) whence, if certain conditions are satisfied, it will be returned to earth for reception at the required locality.

The space wave generally has two components, one of which travels in a very nearly straight line between the transmitting and receiving locations, and the other travels by means of a single reflection from the earth.

The fourth method is a technique that utilizes the ability of a communications satellite orbiting the earth to receive a signal, amplify it, and then transmit it at a different frequency back towards the earth.

The fifth method listed, scatter (fig. 3.2), could be said to be the UHF/SHF equivalent of using skywave transmission for long-distance

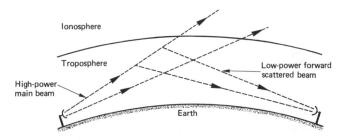

Fig. 3.2
Scatter propagation

HF radio links. The radio energy is directed towards part of the troposphere which forward-scatters the signal towards the receiver. (The scattering region of the troposphere is about 10 km above ground level.)

The radio-frequency spectrum has been subdivided into a number of frequency bands; these are given below.

Frequency band	Classification	Abbreviation
Below 300 Hz	Extremely low	E L F
300 Hz–3 kHz	Infra low	I L F
3 kHz–30 kHz	Very low	V L F
30 kHz–300 kHz	Low	L F
300 kHz–3 MHz	Medium	M F
3 MHz–30 MHz	High	H F
30 MHz–300 MHz	Very high	V H F
300 MHz–3 GHz	Ultra high	U H F
3 GHz–30 GHz	Super high	S H F
30 GHz–300 GHz	Extremely high	E H F
300 GHz–3000 GHz	Tremendously high	T H F

[See Section 6 for *Frequency*]

The surface wave is used for world-wide communications in the low-frequency bands and for broadcasting in the MF band.

The sky wave is used for HF radio communications systems, including long-distance radio-telephony and sound broadcasting.

The space wave is used for sound and TV broadcasting, for multi-channel telephony systems, and for various mobile systems, operating in the VHF, UHF, SHF and higher bands.

Communication satellites are used to carry multi-channel telephony systems, television signals, and data, utilizing UHF and SHF bands.

Scatter systems operate in the UHF and SHF bands to provide multi-channel telephony links.

At UHF and higher frequencies, radio signals can be made extremely directional; antenna systems using large parabolic reflectors produce very narrow radio beams, just as searchlight reflectors produce powerful beams of light. Microwave "dishes" are often used for multi-channel services, both terrestrial and to satellites.

4 Switching System Principles

A switching system of some sort is needed to enable any terminal (e.g. a telephone, a teleprinter, a facsimile unit) to pass information to any other terminal, as selected by the calling customer.

If the network is small, direct links can be provided between each possible pair of terminals and a simple selecting switch installed at each terminal (fig. 4.1). If there are 5 terminals, each must be able to access 4 links, so if there are N terminals there must be a total of $\frac{1}{2}N(N-1)$ links.

A slightly different approach would be to have one link permanently connected to each terminal, always used for calls to that particular terminal (fig. 4.2). Again, each terminal would need a selection switch to choose the distant end wanted for a particular conversation, but the number of links is reduced from 10 to 5 for a 5-terminal system, and to N links for N terminals.

As numbers of terminals and distances increase, this type of arrangement becomes impossibly expensive with today's technologies. A variant of this is however in wide use in radio-telephone networks: all terminals use a single common channel to give the instructions for setting up each call (fig. 4.3). The terminals concerned then both switch to the allocated link or channel for their conversation. The number of links may be reduced substantially by this method; enough need be provided only to carry the traffic generated by the system. But each terminal still needs access to several channels and must have its own selection switch.

So far as telephone networks are concerned it is at present more economical to perform all switching functions at central points (i.e. not to make each terminal do its own switching). This means the provision of only one circuit from the nearest switching point or exchange to each subscriber's terminal (see fig. 4.4). As networks grow and expand into other areas, more switching points are provided, with special circuits between such points to carry the traffic between the areas concerned. The whole network has to be designed to strict transmission parameters so as to ensure that any subscriber on any exchange can converse satisfactorily with all other subscribers, anywhere in the world. Call accounting equipment also has to be provided so that appropriate charges can be levied for all calls made.

The first telephone exchanges were manual and all calls were established by operators. Automatic exchanges using electromechanical relays and switches were however developed very rapidly. Computer-controlled exchanges, with no moving parts at all, have now become common. Most countries in the world now have automatic telephone systems fully interconnected with the rest of the world's systems.

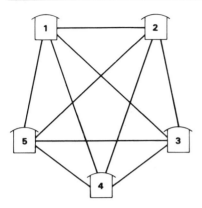

Fig. 4.1
Full interconnection

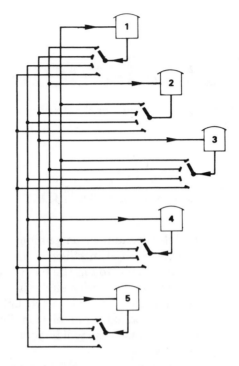

Fig. 4.2
The "one link per terminal" method

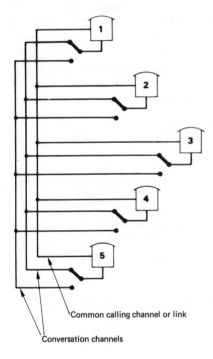

Fig. 4.3
Use of a calling channel

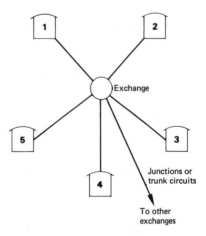

Fig. 4.4
Use of a switching centre: the telephone exchange

Part B
Some Fundamentals

5 The Decibel

In telecommunication engineering, an engineer is concerned with the transmission of intelligence from one point to another, the intelligence being transmitted in the form of electrical signals. A telecommunication system which carries such signals may consist of a number of links in tandem and, certainly, each link will consist of a number of different items, such as transmission lines and amplifiers, also connected in tandem. Each item will introduce a certain loss, or gain, of power into the system.

Consider as a simple example a link of four items:

Item 1 introduces a loss with ratio $\dfrac{\text{Power out}}{\text{Power in}} = \dfrac{1}{2}$

Item 2 introduces a further loss with ratio $\dfrac{\text{Power out}}{\text{Power in}} = \dfrac{1}{50}$

Item 3 introduces a further loss with ratio $\dfrac{\text{Power out}}{\text{Power in}} = \dfrac{1}{10}$

Item 4 introduces a gain with ratio $\dfrac{\text{Power out}}{\text{Power in}} = \dfrac{10\,000}{1}$

Then, if there are no complications in the circuitry, the overall ratio is:

$$\frac{\text{Power out}}{\text{Power in}} = \frac{1}{2} \times \frac{1}{50} \times \frac{1}{10} \times \frac{10\,000}{1} = 10$$

In other words the signal will be ten times as strong at the end as it was at the beginning.

In practice we do not often have such friendly numbers to deal with. The multiplication of a whole chain of numbers like

$$0.001347 \times 0.0418 \times 0.1117 \times 2174.32$$

would, however, be somewhat tedious, and the power ratios we deal with in telecommunications work invariably involve inconveniently large or very small numbers.

To multiply numbers together you do of course have to add together the logarithms of these numbers. Since adding is much easier to do than multiplying, it was decided to concentrate on the logarithms of power ratios rather than the ratios themselves.

This unit, the logarithm of the ratio of in and out powers, is called a Bel, after the Scottish-born American inventor of the telephone, Alexander Graham Bell. The Bel turned out to be too large a unit for normal everyday use so it has been divided into ten to become the decibel or dB.

$$\text{Number of dB} = 10 \log_{10} \frac{\text{Power out}}{\text{Power in}}$$

If half the input power is lost, there is a loss of 3 dB. If an amplifier boosts power by 10 000 times, there is a gain of 40 dB. The decibel is widely used in all telecommunication disciplines, and it is not as complicated as it seems. If, for example, a type of cable introduces 1 dB of loss per kilometre, a 6.0 kilometre length will introduce a total loss of 6.0 dB. If this is followed by two circuits, one giving a loss of 3 dB and the other a gain of 15 dB, the total effect is obtained by plain addition, an overall gain of 6 dB.

The human ear is capable of responding to a wide range of sound intensities and has a sensitivity which varies with change in amplitude in a logarithmic manner. This makes the decibel a convenient unit for use with sound measurement also; it is for example used when measuring the noise of aircraft taking off.

The decibel is not an absolute unit but is only a measure of a power ratio. It is meaningless to say, for example, that an amplifier has an output of 60 dB unless a reference level is quoted or is clearly understood. For example, a 60 dB increase on 1 microwatt gives a power level of 1 watt and a 60 dB increase on 1 watt gives a power level of 1 Megawatt. Here the same 60 dB difference expresses power differences of less than 1 watt in one case and nearly one million watts in the other. It is therefore customary in telecommunication engineering to express power levels as so many decibels above, or below, a clearly understood reference power level, the reference equivalent. This practice makes the decibel a more significant unit and allows it to be used for absolute measurements. The reference level most commonly used is 1 milliwatt, and a larger power, P_1 watts, is said to have a level of

$$+ x \text{ dBm} \quad \text{where } x = 10 \log_{10}\left(\frac{P_1}{1 \times 10^{-3}}\right)$$

and a smaller power, P_2 watts, is said to have a level of

$$- y \text{ dBm} \quad \text{where } y = 10 \log_{10}\left(\frac{1 \times 10^{-3}}{P_2}\right)$$

6 Frequency, Waveforms and Filters

1 Direct and Alternating Current

When an electric current is sent along a wire, energy is dissipated as heat. The amount of heat generated in the wire is energy sent out from the source but lost on the way and not available for use at the distant end; such energy is dependent on the square of the current which is flowing. This means for example that, if the current goes up by a factor of 5, the energy lost on the way goes up by a factor of 25. In some circumstances losses on this scale might well be unacceptable.

Electric power stored in batteries produces a steady voltage which will drive a steady current in one direction, called direct current or d.c., round a circuit and back to the other terminal of the battery.

To transfer large quantities of energy from power stations to distribution points, it is clearly desirable for as low a current as possible to flow, to minimize heat losses. One basic fact about electricity is that the same total amount of power can be sent if the voltage is multiplied by a factor, say 100, and the current divided by the same factor. Heat dissipation on the way (proportional to the square of the current) would then drop to 1/10 000th part of what it was before—or it might be economic to use a thinner and cheaper gauge of wire and accept a reduced heat-loss saving.

Current which can be made to reverse direction at regular intervals is called alternating current, a.c. Alternating current can be fed to a transformer which enables its voltage levels to be changed either up or down to suit particular requirements. The use of a.c. at high voltage enables large amounts of power to be transmitted across a country, without huge heat losses; these high voltages are then transformed down to the voltage levels we use domestically in our houses.

(High d.c. voltages are just as efficient for transferring power with reduced heat losses and are used in some specialized applications. The generation of a.c. is however somewhat simpler than that of d.c., and only a.c. can be fed through transformers to change the supply voltage down from the extremely high voltages used for transmission to levels safe enough to be used in homes.)

One simple form of alternating voltage generator (called an alternator) is shown in fig. 6.1. This is a loop of wire rotated in a magnetic field. The voltage induced in a conductor moving in a magnetic field is proportional to the rate at which the moving wire cuts the magnetic flux. In fig. 6.1*a* the conductors AB and CD cut the magnetic field between the North and South magnetic pole pieces at right angles, so the maximum or peak voltage is induced in the coil. When the wire loop has rotated through one quarter turn to that in fig. 6.1*b*, both conductors are moving parallel to the magnetic field and so are not cutting the flux at all and there is

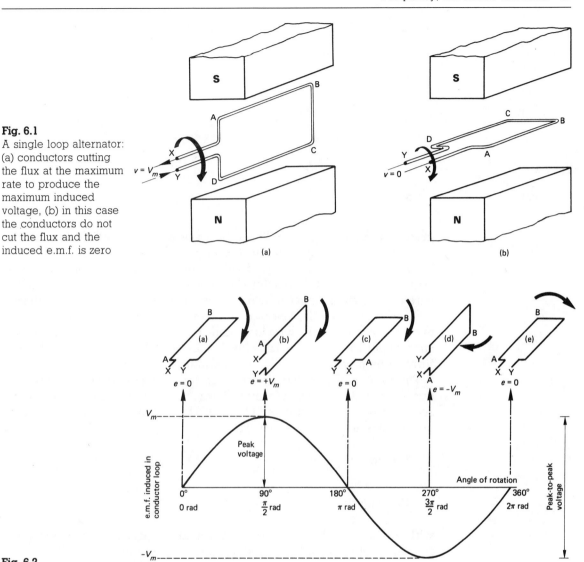

Fig. 6.1
A single loop alternator: (a) conductors cutting the flux at the maximum rate to produce the maximum induced voltage, (b) in this case the conductors do not cut the flux and the induced e.m.f. is zero

Fig. 6.2
One complete cycle of a sinusoidal waveform

zero induced voltage. Figure 6.2 shows how the induced electromotive force or voltage varies over a complete revolution of the loop, providing one cycle of the alternating voltage.

2 The Sinusoidal Waveform

Many waveforms are possible with alternating currents. One of the simplest to produce comes by the rotation of a loop of wire in a uniform magnetic field as described. This is called a sinusoidal waveform and is shown in fig. 6.3.

Between points A and B the current increases from zero to a peak value in the positive direction. Between points B and C the current

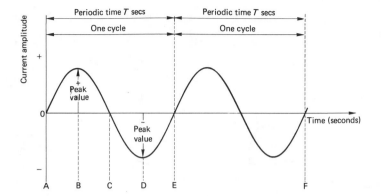

Fig. 6.3
Sinusoidal a.c.
waveform

gradually reduces to zero. Then, between points C and D, the current "increases" to a peak value in the opposite or negative direction, and between points D and E it gradually "reduces" to zero again. This whole sequence from point A to point E represents one complete rotation of the wire loop in the magnetic field, and is called a cycle of a.c. waveform.

Clearly the cycle is repeated between points E and F, representing another rotation of the wire loop, and this waveform is repeated for each subsequent rotation. The time needed, in seconds, for one cycle of waveform to be produced is called the periodic time T of the a.c. waveform.

The number of complete cycles occurring in one second is called the frequency f of the a.c. waveform in hertz (Hz). One hertz is one cycle per second. From fig. 6.4 it should be clear that frequency and periodic time are reciprocals of each other. That is

$$\text{Frequency} = \frac{1}{\text{Periodic time}}$$

$$\text{Periodic time} = \frac{1}{\text{Frequency}}$$

with frequency in hertz and time in seconds. In fig. 6.4, for example, the frequency is 4 Hz and the periodic time is $\frac{1}{4}$ second.

The strength of the current at any instant in time is called the amplitude of the waveform, and the direction of the current (positive or negative) is called the polarity of the current.

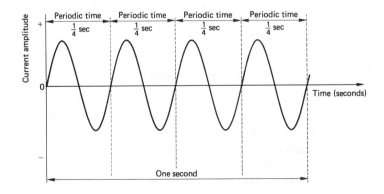

Fig. 6.4
Sinusoidal a.c.
waveform with a
frequency of 4 Hz

It will also be clear from fig. 6.3 and fig. 6.4 that the amplitude reaches a peak value in the positive and negative directions once every cycle.

We have, as one example, already associated the production of a sinusoidal waveform with the rotation of a loop of wire in a magnetic field, and the resulting current plotted against time, as in fig. 6.3 and fig. 6.4. We could also consider the loop as moving through 360° in one rotation, so we could plot the resultant current against angular rotation, as shown in fig. 6.5.

Also, if we consider the energy of an a.c. waveform travelling through space or along a transmission line at a particular velocity, then a certain distance will be travelled in the periodic time for one cycle, as shown in fig. 6.6. We can see now that the a.c. waveform repeats complete cycles over equal distances. The distance representing each cycle is called the wavelength of the a.c. waveform in metres. The Greek letter lambda λ is used as the symbol for wavelength.

In fig. 6.5 the rotation of the loop is shown as a continuously increasing number of degrees. Alternatively, we could consider the start of each rotation as beginning from 0°, so each successive cycle in fact occurs from 0 to 360°, as shown in fig. 6.7. Clearly, at the same point in

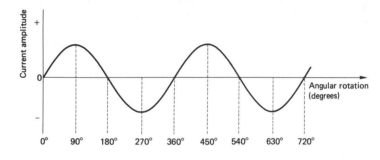

Fig. 6.5
Sinusoidal a.c.
waveform plotted
against angular rotation

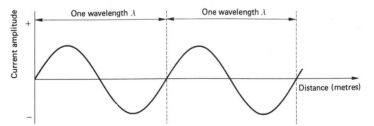

Fig. 6.6
Sinusoidal a.c.
waveform plotted
against distance

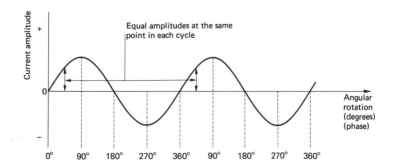

Fig. 6.7
Sinusoidal a.c.
waveform related to
phase

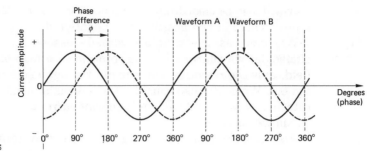

Fig. 6.8
Phase difference
between two waveforms

each cycle, the amplitude of the waveform has the same value. This way
of identifying a particular point in any cycle as a degree of rotation is
called the phase of the a.c. waveform.

In the same way, if two waveforms are identical except for their phase,
then the difference between the two can be expressed as a phase difference, as shown in fig. 6.8. Waveform A is seen to be leading waveform
B by 90°. Put another way, waveform B is lagging waveform A by 90°.

We have seen that an a.c. waveform has a certain energy velocity
(metres per second), with a periodic time (T seconds) for the duration of
each cycle, and with a certain wavelength distance (λ metres) for each
cycle. Now, in general, velocity, distance and time are related by

$$\text{Velocity} = \frac{\text{Distance}}{\text{Time}} \text{ (e.g. metres per second, km/h)}$$

So for any a.c. waveform of wavelength λ and periodic time T,

$$\text{Velocity } v = \frac{\text{Wavelength } \lambda}{\text{Time } T}$$

But it has already been established that

$$\text{Frequency } f(\text{Hz}) = 1/\text{Periodic time } T \text{ (secs)}$$

so that

$$\text{Velocity } v = \text{Wavelength } \lambda \times \text{Frequency } f$$

Therefore, for any a.c. waveform, if two of these properties are known,
the third can be calculated:

$$v = \lambda f \qquad \lambda = \frac{v}{f} \qquad f = \frac{v}{\lambda}$$

The following abbreviations should be noted:

$\text{kHz} = \text{kilohertz} = 10^3 \text{ cycles per second}$
$\text{MHz} = \text{Megahertz} = 10^6 \text{ cycles per second}$
$\text{GHz} = \text{Gigahertz} = 10^9 \text{ cycles per second}$

3 Other Waveforms

Alternating current is not always exactly sinusoidal: waveforms can be
square, triangular, saw-toothed, or in any other repetitive shape, e.g. in
the form of pulses. (See fig. 6.9.)

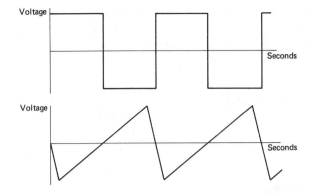

Fig. 6.9
Alternating voltages
with square and
triangular waveforms

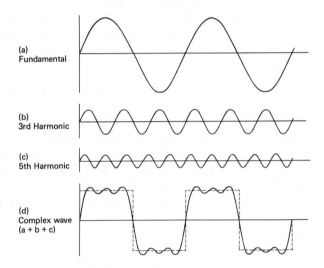

(a)
Fundamental

(b)
3rd Harmonic

(c)
5th Harmonic

Fig. 6.10
Addition of fundamental
sine wave, with third
and fifth harmonics, to
give an approximation
to a square waveform

(d)
Complex wave
(a + b + c)

A mathematician called Fourier proved that any recurrent waveform of frequency F can be resolved into the sum of a number of sinusoidal waveforms having frequencies F, $2F$, $3F$, etc. In other words, any steady signal may be split up into a fundamental (frequency F) and harmonics (multiples of the fundamental frequency F, i.e. $2F$, $3F$, etc). Sounds produced by human voices are nearly always rich in harmonics.

A completely square waveform may, for example, be reconstructed by adding together the fundamental and an infinite series of odd harmonics. In fig. 6.10, a close approximation to a square wave is produced by adding to the fundamental only the 3rd and 5th harmonics.

4 Frequency Ranges

Typical commonly encountered frequencies are

Human voice 100 Hz to 10 000 Hz
Human hearing 20 Hz to 15 000 Hz
Commercial speech 300 to 3400 Hz
Mains electricity 50 or 60 Hz.

If you hear mains-driven electric equipment hum quietly when you switch it on, you are not hearing the electricity; the humming noise is usually due to components being moved or rattled or changing dimensions slightly at the mains frequency.

Radio (broadcast)	1000 kHz
Radio (microwave)	6 GHz
Infra-red rays	100 000 GHz
Visible light	1 000 000 GHz
X-Rays	100 million GHz

5 Filters

It is frequently necessary to be able to separate signals at one frequency from those at other frequencies. The simplest method of accomplishing this is the use of a filter. There are four basic types of filter (fig. 6.11), each of which can be made up in two forms using capacitors and inductors.

1) Low pass—these attenuate all signals at a greater frequency than the cut-off value.
2) High pass—these pass all signals at frequencies higher than the cut-off value and attenuate lower frequency signals.
3) Band stop—these attenuate signals at frequencies within a specified band and pass signals at frequencies above and below this band.
4) Band pass—these pass all signals at frequencies within a specified band and attenuate signals at frequencies above and below the band.

Fig. 6.11
Basic filters

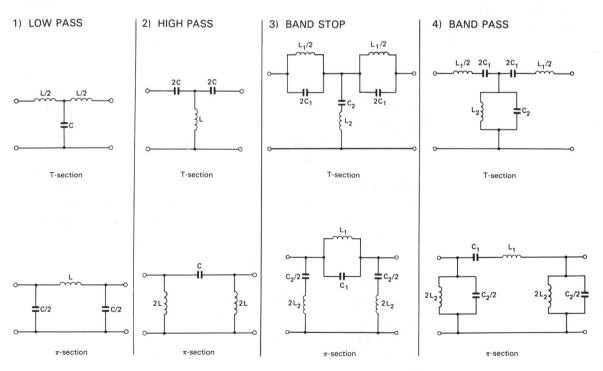

7 Voice Frequencies

When we speak, our vocal cords vibrate and the resulting sounds, which are rich in harmonics but of almost constant pitch, are carried to the cavities in the mouth, throat and nose. Here the sounds are given some of the characteristics of the desired speech by the emphasizing of some of the harmonics contained in the sound waveforms and the suppression of others. Sounds produced in this way are the vowels, a, e, i, o and u, and contain a relatively large amount of sound energy. Consonants are made with the lips, tongue and teeth and contain much smaller amounts of energy and often include some relatively high frequencies.

The sounds produced in speech contain frequencies which lie within the frequency band 100–10 000 Hz. The pitch of the voice is determined by the fundamental frequency of the vocal cords and is about 200–1000 Hz for women and about 100–500 Hz for men.

The power content of speech is small, a good average being of the order of 10–20 microwatt. However, this power is not evenly distributed over the speech-frequency range, most of the power being contained at frequencies in the region of 500 Hz for men and 800 Hz for women.

The notes produced by musical instruments occupy a much larger frequency band than that occupied by speech. Some instruments, such as the organ and the drum, have a fundamental frequency of 50 Hz or less, while many others, such as the violin and clarinet, can produce notes having a significant harmonic content at frequencies of up to 15 000 Hz. The power content of music can be quite large. A sizeable orchestra may generate a peak power somewhere in the region of 90–100 watts while a bass drum well thumped may produce a peak power of about 24 watts.

When sound waves are incident upon the ear they cause the ear drum to vibrate. Coupled to the ear drum are three small bones which transfer the vibration to a fluid contained within a part of the inner ear known as the cochlea. Inside the cochlea are a number of hair cells, and the nerve fibres of these are activated by vibration of the fluid. Activation of these nerve fibres causes them to send signals, in the form of minute electric currents, to the brain where they are interpreted as sound.

The ear can only hear sounds whose intensity lies within certain limits; if a sound is too quiet it is not heard and, conversely, if a sound is too loud it is felt rather than heard and causes discomfort or even pain. The minimum sound intensity that can be detected by the ear is known as the "threshold of hearing or audibility" and the sound intensity that just produces a feeling of discomfort is known as the "threshold of feeling". The ear is not, however, equally sensitive at all frequencies, as shown in fig. 7.1. In this diagram, curves have been plotted showing how the thresholds of audibility and feeling vary with frequency for an average person.

It can be seen that the frequency range over which the average human

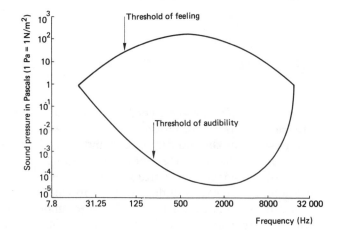

Fig. 7.1
The thresholds of
audibility and feeling

ear is capable of responding is approximately 30—16 500 Hz, but this range varies considerably with the individual. The ear is most sensitive in the region of 1000 to 2000 Hz and becomes rapidly less sensitive as the upper and lower limits of audibility are approached. The limits of audibility are clearly determined not only by the frequency of the sound but also by its intensity. At the upper and lower limits of audibility, the thresholds of audibility and feeling coincide and it becomes difficult for an observer to distinguish between hearing and feeling a sound. As human beings get older they usually become incapable of hearing high-frequency notes, unless the power level is made exceedingly high, a situation eased by the advent of low-priced high-fidelity sound reproduction systems.

In an ideal telecommunications system, all the frequencies present in a speech or music waveform would be converted into electrical signals, transmitted over the communication system, and then reproduced as sound at the distant end. In practice, this is rarely the case, for two reasons. Firstly, for economic reasons the devices used in circuits that carry speech and music signals have a limited bandwidth; secondly, particularly for the longer-distance routes, a number of circuits are often transmitted over a single telecommunication system and this practice provides a further limitation of bandwidth. It is thus desirable to have some idea of the effect on the ear when it is responding to a sound wave-form, the frequency components of which have amplitude relationships differing from those existing in the original sound.

By international agreement the audio-frequency band for a "commercial quality" speech circuit routed over a multi-channel system is restricted to 300—3400 Hz. This means that both the lower and upper frequencies contained in the average speech waveform are not transmitted. To the ear, the pitch of a complex, repetitive sound waveform is the pitch corresponding to the frequency difference between the harmonics contained in the waveform, i.e. the pitch is that of the fundamental frequency. Hence, even though the fundamental frequency itself may have been suppressed, the pitch of the sound heard by the listener is the same as the pitch of the original sound. However, much of the power contained in the original sound is lost. Suppression of all frequencies above 3400 Hz reduces the quality of the sound but does not affect its intelligibility. Since the function of a telephone system is to transmit intelligible speech, the loss of quality can be tolerated; sufficient quality remains to allow a speaker's voice to be recognized.

Music is, however, badly distorted when transmitted over a normal telephone line because both low and high frequency notes are lost.

If a telecommunications link is required to carry a television signal, an even wider bandwidth needs to be transmitted without distortion. A rough rule of thumb for the bandwidths needed for acceptable transmission is:

Voice 4 kHz   Music 10 kHz–15 kHz   Colour TV 8 MHz

8 Attenuation and Noise

As an information signal travels along a line its amplitude or power is progressively reduced because of losses in the line. These losses, called attenuation, are of two types:

1) Losses due to heat dissipation caused by the resistance of the conductors and the insulation resistance between the conductors.
2) Dielectric losses which affect alternating current (or a.c.) only and are dependent on the dimensions and type of insulant between the conductors themselves and between conductors and earth.

Attenuation generally increases as the frequency of the information signal increases; this variation is called attenuation distortion.

Figure 8.1 shows how the attenuation of ordinary audio-frequency cable pairs of different gauges (diameters of 0.9 and 0.63 mm) varies with frequency. As you would expect, the thicker the wire the lower the attenuation. A telephone cable serving a subscriber many kilometres from the exchange may therefore have to be made up of heavier gauge conductors than those needed to serve phones in an office just next door to the exchange. It is not uncommon for there to be a loss of 10 dB between a subscriber and the subscriber's own exchange; this means that only 1/10th of the transmitted power of the original voice signal reaches the local exchange. Similar losses could be experienced at the other end also, so even an own-exchange call can often mean that only 1/100th of the original power is available: a 20 dB loss. Even with 30 dB of attenuation (1/1000th of the power), speech is usually still possible, so long as there isn't too much noise on the line.

The attenuation suffered by a signal passing along an audio-frequency cable like this continues to increase at frequencies above those plotted in the figure and could be as high as 30 dB/km at a typical carrier frequency of 1 MHz. For comparison, the attenuation/frequency

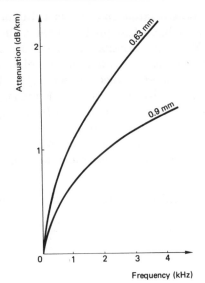

Fig. 8.1
Attenuation/frequency
characteristics of two
typical audio-frequency
cables

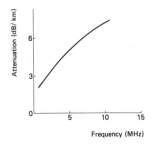

Fig. 8.2
Attenuation/frequency
characteristics of a
coaxial pair

characteristics for a coaxial cable, designed to be used for carrier telephony at high frequencies, are given in fig. 8.2.

In any telecommunication system, whether using line or radio links, there is unwanted electrical energy present as well as that of the wanted information signal. This unwanted electrical energy is generally called noise and arises from a number of different sources, which will now be considered very briefly.

1) *Resistor noise*   A conductor is designed to carry current with minimum opposition, consistent with size and cost.

A resistor is a component designed to have a particular opposition to the flow of electrical current in a particular circuit. This opposition is called resistance in d.c. circuits, but in a.c. circuits the term impedance is used because of other, frequency-dependent, factors. In either case the unit used is the ohm (symbol Ω).

An electric current is produced by the movement of electrons dislodged by an externally applied voltage from the outer shells of the atoms making up the conductor material or resistor material. The movement or agitation of atoms in conductors and resistors is somewhat random, and is determined by the temperature of the conductor or resistor. The random movement of electrons brought about by thermal agitation of atoms tends to have increased energy as temperature increases.

This random movement of atoms gives rise to an unwanted electrical voltage which is called resistor noise, circuit noise, Johnson noise or thermal noise. This unwanted signal spreads over a wide range of frequencies, and the noise present in a given bandwidth required for a particular information signal is very important. This is the noise temperature of the resistor or conductor, measured in the Kelvin temperature scale, which has its zero point at $-273°$ Centigrade. This is the temperature at which the random movement or agitation of atoms in conducting or resistive materials ceases, so unwanted noise voltages are therefore zero.

2) *Shot noise*   This is the name given to noise generated in active devices (energy sources), such as valves and transistors, by the random varying velocity of electron movement under the influence of externally applied potentials or voltages at appropriate terminals or electrodes.

3) *Partition noise*   This occurs in multi-electrode active devices such as transistors and valves and is due to the total current being divided between the various electrodes.

4) *Fluctuation noise*   This can be natural (electric thunderstorms, etc.) or man-made (car ignition systems, electrical apparatus, etc.) and again spreads over a wide range of frequencies. Such noise can be picked up by active devices and conductors forming transmission lines.

5) *Static*   This is the name given to noise encountered in the free-space transmission paths of radio links, and is due mainly to ionospheric storms causing fluctuations of the earth's magnetic field. This form of noise is affected by the rotation of the sun and by the sunspot activity that prevails.

6) *Cosmic or Galactic noise*   This type of noise is also most troublesome to radio links, and is mainly due to nuclear disturbances in all the galaxies of the universe.

7) *Crosstalk*   In multi-pair cables there is capacitive and inductive coupling between different pairs which produces an unwanted noise signal on any pair because signals are transmitted from other pairs. This is called crosstalk and can be reduced to some extent by twisting the conductors of each pair, or by changing the relative positions of pairs along the cable during manufacture, or by balancing the pairs over a particular route after installation.

8) *1/f noise*   Fluctuations in the conductivity of the semiconductor material produce a noise source which is inversely proportional to frequency. This type of noise is also known as current noise, excess noise, or flicker noise.

In any telecommunications system, therefore, there will be a certain level of noise power arising from all or some of the sources described, with the noise power generally being of a reasonably steady mean level, except for some noise arising from impulsive sources such as car ignition systems and lightning. Noise which has a sensibly constant mean level over a particular frequency bandwidth is generally called white noise.

The presence or absence of unwanted noise on a circuit is one of the ways in which the quality of a circuit can be described. The received level is measured when a wanted signal of specified power level is present. This is compared with the received level measured when there is no signal present. This enables the ratio

$$\frac{\text{Signal plus Noise}}{\text{Noise}}$$

to be calculated. It is usually expressed in decibels and called the signal-to-noise ratio. The higher the ratio, the higher the quality of the circuit.

9 Amplitude Modulation

We have seen from section 7 that a bandwidth of from 300 Hz to 3400 Hz is required for the transmission of commercial quality speech.

To economize on cable it is desirable to be able to transmit more than one conversation over a single pair of wires. If several conversation signals were all connected together at one end of a line, it would not be possible to separate them at the distant end since each conversation would be occupying the same frequency spectrum of 300 Hz to 3400 Hz. Amplitude modulation (AM) plus frequency division multiplexing (FDM) is one way of solving this problem. Each conversation is shifted to a different part of the frequency spectrum by using a high-frequency waveform to "carry" each individual speech signal. These high frequencies are called carrier frequencies.

Amplitude modulation is the process of varying the amplitude of the sinusoidal carrier wave by the amplitude of the modulating signal, and is illustrated in fig. 9.1.

The unmodulated carrier wave has a constant peak value and a higher frequency than the modulating signal but, when the modulating signal is applied, the peak value of the carrier varies in accordance with the instantaneous value of the modulating signal, and the outline wave shape or "envelope" of the modulated wave's peak values is the same as the original modulating signal wave shape. The modulating signal waveform has been superimposed on the carrier wave.

It can be shown by mathematical analysis that, when a sinusoidal carrier wave of frequency f_c Hz is amplitude-modulated by a sinusoidal modulating signal of frequency f_m Hz, then the modulated carrier wave contains three frequencies.

One is the original carrier frequency, f_c Hz.

The second is the sum of carrier and modulating signal frequencies, $(f_c + f_m)$ Hz.

The third is the difference between carrier and modulating signal frequencies, $(f_c - f_m)$ Hz.

This is illustrated in fig. 9.2.

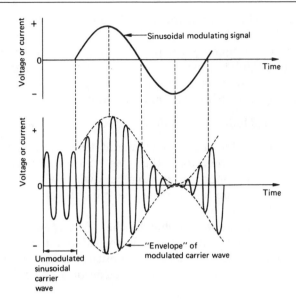

Fig. 9.1
Amplitude-modulated
carrier wave

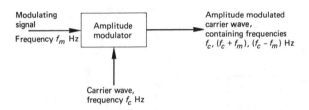

Fig. 9.2
Principle of amplitude
modulation

It should be noted that two of these frequencies are new, being produced by the amplitude-modulation process, and are called side-frequencies.

The sum of carrier and modulating signal frequencies is called the upper side-frequency. The difference between carrier and modulating signal frequencies is called the lower side-frequency. This is illustrated in the frequency spectrum diagram of fig. 9.3.

The bandwidth of the modulated carrier wave is

$$(f_c + f_m) - (f_c - f_m) = 2f_m$$

i.e. double the modulating signal frequency.

Fig. 9.3
Frequency spectrum of
an amplitude-modulated
wave for
single-frequency
modulating signal

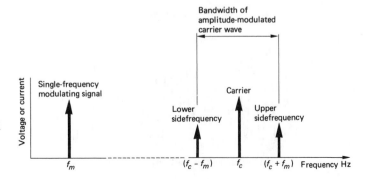

When the modulating signal consists of a band of frequencies, as already seen for commercial speech and music for example, then each individual frequency will produce upper and lower side-frequencies about the unmodulated carrier frequency, and so upper and lower sidebands are obtained. This is illustrated in fig. 9.4.

The bandwidth of the modulated carrier wave is

$$(f_c + 3400) - (f_c - 3400) = 6800\,\text{Hz}$$

which is double the highest modulating signal frequency.

It follows therefore that, as the modulating signal bandwidth increases, the modulated wave bandwidth also increases, and the transmission system used must be capable of handling this bandwidth throughout.

Figure 9.5 shows graphically the effect of adding together two signals, representing an upper and a lower sideband. The envelope of the resultant waveform is typical of the output from various types of balanced modulator as used in the production of amplitude-modulated waves.

Fig. 9.4
Frequency spectrum of an amplitude-modulated wave for commercial speech modulating signal

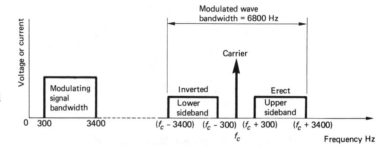

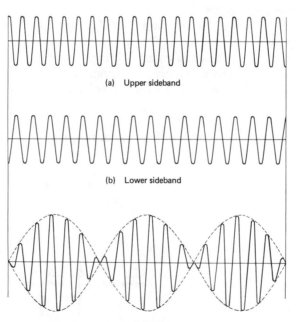

(a) Upper sideband

(b) Lower sideband

Fig. 9.5
Addition of two sidebands (without carrier)

(c) Upper and lower sideband present simultaneously

The complete amplitude-modulated wave band of lower sideband plus carrier plus upper sideband shown in fig. 9.4 takes up more frequency bandwidth than is really necessary to transmit the information signal since all the information is carried by either one of the sidebands alone. The carrier component is of constant amplitude and frequency so does not carry any of the information signal at all. It is possible by using special equipment to suppress both the carrier and one sideband and to transmit just the other sideband with no loss of information. This method of working is called single sideband working (SSB), or single-sideband suppressed-carrier working. This method is more costly and complex than transmitting the wider band carrying the two sidebands plus the carrier and is not used for domestic radio broadcasting, but it is used for some long-distance radio telephony systems and for multi-channel carrier systems used in national telephone networks.

10 Frequency Modulation

Another method of superimposing information signals on to a carrier signal is frequency modulation in which the modulating signal varies the frequency of a carrier wave. This has a number of advantages over amplitude modulation. Frequency modulation is used for sound broadcasting in the VHF band, for the sound signal of 625-line television broadcasting, for some mobile systems, and for multi-channel telephony systems operating in the UHF band. Frequency modulation is therefore used for all analog communications via satellites. The price which must be paid for some of the advantages of frequency modulation over double sideband amplitude modulation is a wider bandwidth requirement.

When a sinusoidal carrier wave is frequency modulated, its instantaneous frequency is caused to vary in accordance with the characteristics of the modulating signal. The modulated carrier frequency must vary either side of its nominal unmodulated frequency a number of times per second equal to the modulating frequency. The magnitude of the variation—known as the frequency deviation—is proportional to the amplitude of the modulating signal voltage.

The concept of frequency modulation can perhaps best be understood by considering a modulating signal of rectangular waveform, such as the waveform shown in fig. 10.1. Suppose the unmodulated carrier frequency is 3 MHz. The periodic time of the carrier voltage is $\frac{1}{3}$ μs and so three complete cycles of the unmodulated carrier wave will occur in 1 μs. When, after 1 μs, the voltage of the modulating signal increases to +1 V, the instantaneous carrier frequency increases to 4 MHz. Hence in the time interval 1 μs to 2 μs, there are four complete cycles of the carrier voltage. After 2 μs the modulating signal voltage returns to 0 V and the instantaneous carrier frequency falls to its original 3 MHz. During the time interval 3 μs to 4 μs the modulating signal voltage is −1 V and the

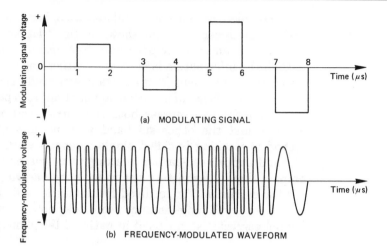

Fig. 10.1
A frequency-modulated
wave

carrier frequency is reduced to 2 MHz; this means that two cycles of the carrier voltage occur in this period of time.

When, after 4 μs, the modulating voltage is again 0 V, the instantaneous carrier frequency is restored to 3 MHz. At $t = 5$ μs the modulating voltage is $+2$ V and, since frequency deviation is proportional to signal amplitude, the carrier frequency is deviated by 2 MHz to a new value of 5 MHz. Similarly, when the modulating voltage is -2 V, the deviated carrier frequency is 1 MHz. At all times the amplitude of the frequency modulated carrier wave is constant at 1 V, and this means that the modulating process does not increase the power content of the carrier wave.

When the modulating signal is of sinusoidal waveform, the frequency of the modulated carrier wave will vary sinusoidally; this is illustrated by fig. 10.2.

The frequency deviation of a frequency-modulated carrier wave is proportional to the amplitude of the modulating signal voltage. There is no inherent maximum value to the frequency deviation that can be obtained in a frequency-modulation system; this should be compared with amplitude modulation where the maximum amplitude deviation possible corresponds to 100% modulation, i.e. reduction of the minimum amplitude of the "envelope" in fig. 9.1 to zero.

The modulation of frequency may also be regarded as a modulation of the phase of the carrier wave. In phase modulation the phase is varied in accordance with the modulating signal. When the modulating signal is a single tone, frequency and phase modulated signals are indistinguishable. However, when the modulating signal frequency is varied or the modulating signal is a complex (multi-frequency) signal, they are different since in phase modulation the effective frequency deviation depends on the frequency of the modulating signal. Phase modulation can be obtained by applying a weighting network to the input of a frequency modulator. Phase modulation is considered to be advantageous in some situations due to the larger deviation at higher modulating frequency effectively increasing the signal-to-noise ratio.

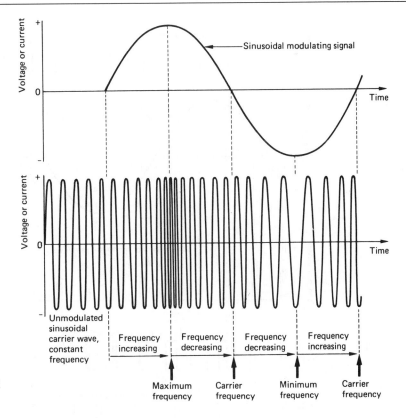

Fig. 10.2
A frequency-modulated
carrier wave

11 Pulse Modulation

Another method of conveying information is by means of pulses of voltage or current.

With pulse modulation the carrier wave is not sinusoidal, but consists of repeated rectangular pulses. The amplitude, width or position of the pulses can be altered by the information signal, as illustrated in fig. 11.1.

Pulse amplitude modulation is that form of modulation in which the amplitude of the pulse carrier is varied in accordance with some characteristic, normally the amplitude, of the modulating signal.

Pulse width modulation is that form of modulation in which the duration of a pulse is varied in accordance with some characteristic of the modulating signal. Sometimes pulse width modulation is called pulse duration modulation or pulse length modulation.

Pulse position modulation is that form of modulation in which the positions in time of the pulses are varied in accordance with some characteristic of the modulating signal without a modification of pulse width.

Pulse code modulation is a special form of pulse modulation and is

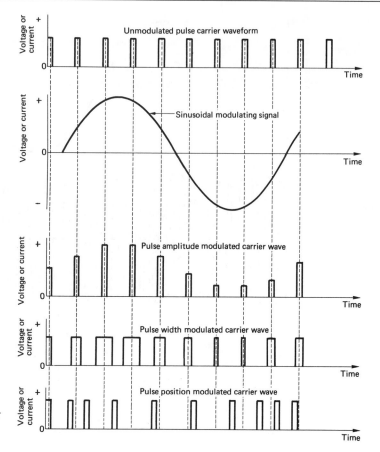

Fig. 11.1
Pulse-modulated carrier waves

dealt with in some detail in section 13; PCM can be considered to be the sampling of the modulating analog signal and the production of pulse amplitude modulated signals, i.e. pulse signals whose amplitudes are directly proportional to the amplitude of the original signal at the successive instants at which samples are measured. Each of these samples is then "quantized" or measured against a built-in scale of amplitudes (from 0 to 256 in modern systems); the number of this scale which represents the amplitude of each particular sample is then transmitted as a binary coded signal, i.e. a coded group of 8 bits (see section 13), to indicate the amplitude level ($2^8 = 256$). The signal in the line is therefore a series of pulses representing numbers. At the receiving end these numbers are decoded and an appropriate analog signal recreated.

Many adaptive forms of transmission of voice information by digital pulses have been proposed. These save bandwidth by transmitting only changes in the modulating signal.

The analog signal is scanned, typically 32 000 times per second (but lower scanning rates can be used in some circumstances). If the signal's instantaneous value is greater than it was at the previous scan, a digit 1 signal is transmitted; if less, a digit 0. Although Delta Modulation (DM) as this is called usually samples speech waveforms more frequently than PCM (PCM only 8000 times per second), each DM signal is only a single digit, 0 or 1, whereas each PCM sample needs a "PCM word" of

8 bits to indicate the quantizing level. DM does therefore have some significant advantages over PCM:

a) It provides greater channel capacity for a given bit-rate, resulting in higher pair-gain and lower per-channel costs.

b) It does not inherently require synchronization, as PCM does.

c) It is more tolerant of system noise.

A form of adaptive PCM, Adaptive Differential Pulse Code Modulation (ADPCM), has now been agreed internationally for transmission of voice signals at 32 000 bits per second. This is specified in CCITT Recommendation G.732.

Because adaptive coding relies on sending changes to the signal only, it is not suitable for the transmission of data signals. Many users send data over voice circuits by modulating the voice frequency with the data signal using modulators and demodulators (Modems); this is dealt with in section 34. For circuits used in such a way, ADPCM would not be suitable.

12 Analog versus Digital

1 Two-state Devices

Much of present-day electronic and telecommunication equipment is still analog in nature. This means that the signals to be handled, processed or transmitted are represented by voltages whose amplitude and/or frequency vary continuously with time; thus, in a telephone system, the transmitted signals are replicas of the speech waveforms. Many examples of analog equipment are well known; for example, the radio and television receivers to be found in the majority of homes.

Digital signals are not continuous in nature but consist of discrete pulses of voltage or current which represent the information to be processed. Digital voltages can vary only in discrete steps; normally only two voltage levels are used—one of which is zero—so that two-state devices can be employed. A two-state device is one which has only two stable states; so that it is either ON or it is OFF. Examples of two-state devices are: a lamp which is either glowing visibly or it is not; a buzzer which is either producing an audible sound or not; or an electrical switch which either completes an electrical circuit or breaks it.

The advantages to be gained from the use of digital techniques instead of analog methods arise largely from the use of just the two voltage levels. Digital circuitry, mainly integrated circuitry in modern systems, operates by switching transistors ON and OFF and does not need to produce or to detect precise values of voltage and/or current at particular points in an equipment or system. Because of this it is easier and cheaper to mass-produce digital circuitry. Also, the binary nature of the signals makes it much easier to obtain consistently a required operating performance from a large number of circuits. Digital circuits are generally more reliable than analog circuits because faults will not often occur through variations in performance caused by changing values of components, misaligned coils, and so on. Again, the effects of noise and interference are very much reduced in a digital system since the digital pulses can always be regenerated and made like new whenever their wave shape is becoming distorted to the point where errors are likely. This is not possible in an analog system where the effect of unwanted noise and interference signals is to permanently degrade the signal.

There are two main reasons why the application of digital techniques to both electronics and telecommunications has been fairly limited in scope until recent years. First, digital circuitry was, in the main, not economic until integrated circuits became freely available, and, secondly, the transmission of digital signals requires the provision of circuits with a very wide bandwidth. Some digital circuits and equipments have, of course, been available since pre-integrated circuit days but their scope and application were very limited.

Although digital techniques clearly have significant advantages we live in a world where we usually count in units of 10 and where phenomena such as sound and light are basically analog in nature, i.e. continuously variable waveforms. How then can we convert from a decimal counting system and analog waveforms into this digital world?

2 Numbering Systems, Coding and Digital Efficiency

Human beings usually count in tens. No doubt we use a decadic numbering system because we happen to have a total of 10 fingers and thumbs. It is very much simpler, however, to make an electronic device operate at only two levels (On or Off, Yes or No, 1 or 0) than to make it so that it can select any one of ten different levels. This is of course one of the main reasons why digital transmission and switching equipment are now becoming cheaper and more efficient than the analog equivalents.

A two-level numbering scheme is called a BINARY (or base 2) system. Just as our decadic or base 10 system uses zero and figures up to 9 (i.e. 10 minus 1), a binary or base 2 system uses 0 (zero) and the figure 1 only (i.e. 2 minus 1).

As a simple exercise let us consider the number 1985. In decadic this means

$$(1 \times 10^3) + (9 \times 10^2) + (8 \times 10^1) + (5 \times 10^0)$$

We can change this into straightforward binary in two ways: either on a one-decimal-digit-at-a-time basis called BINARY CODED DECIMAL or taking the number as a whole.

Binary coded decimal for 1985 is therefore (see table, p. 33)

    0001 1001 1000 0101

To convert a decimal number as a whole to binary, we divide the

BCD: Binary Coded Decimal

Decimal	Binary Coded Decimal
0	0000
1	0001
2	0010
3	0011
4	0100
5	0101
6	0110
7	0111
8	1000
9	1001

decimal number by 2 and then keep on dividing by 2 until there is nothing left. For example,

$$1985 \div 2 = 992 \quad \text{remainder 1}$$
$$992 \div 2 = 496 \quad \text{remainder 0}$$
$$496 \div 2 = 248 \quad \text{remainder 0}$$
$$248 \div 2 = 124 \quad \text{remainder 0}$$
$$124 \div 2 = 62 \quad \text{remainder 0}$$
$$62 \div 2 = 31 \quad \text{remainder 0}$$
$$31 \div 2 = 15 \quad \text{remainder 1}$$
$$15 \div 2 = 7 \quad \text{remainder 1}$$
$$7 \div 2 = 3 \quad \text{remainder 1}$$
$$3 \div 2 = 1 \quad \text{remainder 1}$$
$$1 \div 2 = 0 \quad \text{remainder 1}$$

Therefore, the binary code equivalent of decimal 1985 is

11111000001

i.e. $(1 \times 2^{10}) + (1 \times 2^9) + (1 \times 2^8) + (1 \times 2^7) + (1 \times 2^6) + (1 \times 2^0)$

i.e. Decimal $1024 + 512 + 256 + 128 + 64 + 1$

It will be seen that there are only 11 bits in this binary code for 1985, compared with 16 in the binary coded decimal version of 1985.

Although electronic equipment operates on a binary basis, a Yes–No basis, and we use binary codes for many purposes, computers themselves do not always use straightforward binary for their own calculations. There are several good reasons for this.

Digital efficiency and economics If we use four binary digits (bits) to represent each decimal digit, it is seen from the BCD table that the combinations 1010, 1011, 1100, 1101, 1110 and 1111 have not been used. This means that, since computers have to be designed to handle all the digits input to them, they will be capable of *handling* all 16 of the possible combinations of each 4-bit word but are *using* only 10 of them; they will not be using these 6 combinations at all. In other words, they can at best be only

$$\frac{10}{16} \times 100 \text{ or } 62.5\% \text{ efficient}$$

There are basically two common ways round this in order to increase the efficiency of a computer's internal workings:

a) Restricting the machine to only 3 bits per decimal number, i.e. to use an octal or base 8 code, *or*

b) Making good use of all 4 bits per decimal number by using numbers on a Hexadecimal base, i.e. on base 16.

If 1985 is coded on an OCTAL basis we get the following figures:

$$
\begin{aligned}
1985 = \quad & 3 \times 8^3 \quad (=\text{decimal } 1536) \\
+ & 7 \times 8^2 \quad (=\text{decimal } 448) \\
+ & 0 \times 8^2 \quad (=\text{decimal } 0) \\
+ & 1 \times 8^0 \quad (=\text{decimal } 1)
\end{aligned}
$$

i.e. 3, 7, 0, 1, or in binary coded octal

011   100   000   001

so a total of 12 bits are needed, compared with 16 bits in ordinary binary coded decimal.

In HEXADECIMAL or base 16 code, the extra numbers are given letter codes, as in the table.

Hexadecimal Code

Decimal	Hexadecimal	Binary Hex
0	0	0000
1	1	0001
2	2	0010
3	3	0011
4	4	0100
5	5	0101
6	6	0110
7	7	0111
8	8	1000
9	9	1001
10	A	1010
11	B	1011
12	C	1100
13	D	1101
14	E	1110
15	F	1111

1985 when coded into hex is therefore

$$
\begin{aligned}
& (7 \times 16^2) \quad (=\text{decimal } 1792) \\
+ & (12 \times 16^1) \quad (=\text{decimal } 192) \\
+ & (\ 1 \times 16^0) \quad (=\text{decimal } 1) \\
& \qquad\qquad\qquad (=\text{decimal } 1985)
\end{aligned}
$$

i.e. 7C1, or in binary coded hexadecimal

0111   1100   0001

—again using only 12 bits instead of the 16 bits of ordinary BCD.

To minimize confusion it is not usual for computers to print out answers in octal or hex code; it would be all too easy for mistakes to be made by humans if hex figures were used and mistaken for ordinary decimals: e.g. $707 in hex represents $1799 in ordinary decimals and $707 in octal represents $455.

Some computers use 7-bit words for decimal numbers. This scheme gives each number two parts: the first two bits, the "bi" part, shows whether the number is less than 5 or not:

01 = less than 5   10 = 5 or more

The other 5 bits include only one single 1 code, all the rest are 0s, so the whole representation always has exactly two 1s in it for each digit. If more or less than two are received there has been an error, possibly in transmission, so equipment using this BI-QUINARY code can readily initiate error detection and correction.

Bi-quinary Code

Decimal	Bi-quinary
0	01 00001
1	01 00010
2	01 00100
3	01 01000
4	01 10000
5	10 00001
6	10 00010
7	10 00100
8	10 01000
9	10 10000

3 Analog-to-Digital Conversion

Conversion of analog-type signals such as sound waves to digital format presents more difficulties. The system which is now used for digital telephony was invented in the 1930s by an English engineer, Dr Alec Reeves, working in the ITT Laboratories in Paris. At that time, the technology was not available to make this "pulse code modulation" or PCM system commercially viable for telephony. It has, in fact, taken forty years to reach international agreement on standards to be used for PCM and even now there are two different standards, one basically European, one American.

In summary, the speech signal is sampled 8000 times per second; the amplitude of the signal at the instant of sampling is compared with a built-in series of levels called quantizing intervals; the number of the interval within which the sample falls is sent in an 8-digit binary format to the distant end. At the distant end this binary number (representing decimal figures between −128 to +128) is fed into a decoder and a signal set with amplitude almost exactly the same as the original sample is recreated locally. Basically similar sampling and "quantizing" procedures are followed for the conversion of other electrical waveforms into digital signals, but since music uses higher frequencies than voice a higher sampling rate is needed. Compact discs for example normally utilize a sampling rate of 41 100 per second, with a 16-bit signal to represent each quantized sample.

13 Pulse Code Modulation

In a PCM system the analog signal is sampled at regular intervals to produce a pulse amplitude modulated waveform.

If an analog signal is sampled regularly using a sampling rate of at least twice the highest frequency of the signal, the samples are found to be adequate to allow the recreation of the original voice signal with sufficient accuracy for all purposes. Sampling is done by feeding the analog signal to a circuit with a gate which only opens for the duration of the sampling pulse. The output is a pulse amplitude modulated (PAM) signal (fig. 13.1).

Figures 13.2 and 13.3 show the effect of taking these samples from sound signals of different voice frequencies. Here it has to be remembered that, although the commercial voice band goes up to 3400 Hz,

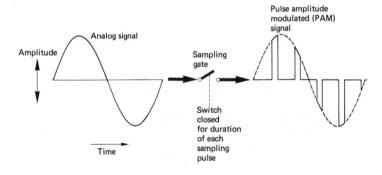

Fig. 13.1
Principle of pulse amplitude modulation (*Courtesy*: An Introduction to Digital Telephony, ITT)

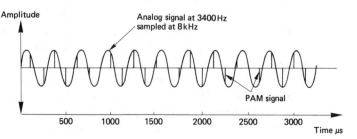

Fig. 13.2
Sampling of the highest voice frequency to be transmitted (*Courtesy*: ITT)

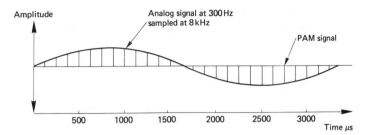

Fig. 13.3
Sampling of the lowest voice frequency to be transmitted (*Courtesy:* ITT)

almost all the power of human speech is at much lower frequencies (e.g. at around 500 Hz for male voices), so several samples are taken during each cycle, enough to enable the original analog signal to be reconstituted with a fair degree of accuracy.

The total amplitude range that the signal may occupy is divided into a number of levels, each of which is allocated a number: there are 256 different levels in internationally recommended PCM. Here it should be noted that there are two different PCM encoding laws in use in the world, called μ-law coding (developed and used in America) and A-law coding (developed and used in Europe). These two laws give different quantization values for a signal of a given amplitude so that PCM channels to these two standards cannot directly interwork unless special interfacing equipment is provided.

The variable-amplitude PAM signal is then compared with the instantaneous value of the appropriate range of levels which are called quantizing intervals. The signal is assigned the value of the interval in which it falls and the number of this value is then encoded into an 8-digit binary code. (This gives 2^8 or 256 possible levels, 128 each side of zero.) Each 8-digit code describing the amplitude level of the sample is known as a PCM word. Since the sampling rate is 8000 times per second and each sample has 8 digits, there are 64 000 bits per second for a single PCM channel.

Figure 13.4 shows how this quantization process works for a complex analog signal. In this figure only 8 sampling levels are shown, for clarity;

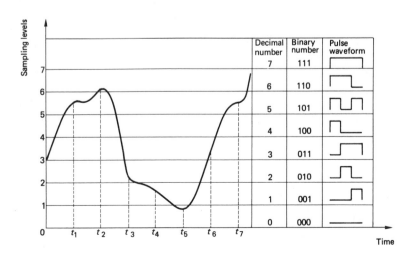

Fig. 13.4
Quantization of a signal

these 8 levels require only 3 binary digits against the 8 binary digits needed to distinguish between the 256 levels used in practical PCM systems.

The signal waveform is sampled at time instants t_1, t_2, t_3, etc. At time t_1 the instantaneous signal amplitude is in between levels 5 and 6 but since it is nearer to level 6, it is approximated to this level. At instant t_2 the signal voltage is slightly greater than level 6 but is again rounded off to that value. Similarly the sample taken at t_3 is represented by level 2, the t_4 sample by level 2, the t_5 sample by level 1, and so on. The binary pulse train which would be transmitted to represent this signal is shown in fig. 13.5. A space, equal in time duration to one binary pulse, has been left in between each binary number in which synchronization information can be transmitted.

Fig. 13.5
Binary pulse train representing the signal shown in Fig. 13.4

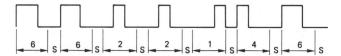

A PCM system transmits signal information in digital form. The quantization process will result in some error at the receiving end of the system when the analog signal is reconstituted. The error appears in the form of quantization noise and can be reduced only by increasing the number of sampling levels. Unfortunately, this would increase the number of binary digits required to signal the sampling level numbers and this, in turn, means that the bandwidth which must be provided would be wider. Practical systems have to accept a compromise solution here. With an 8 binary digits scheme, the quantizing noise is generally considered acceptable. For the receiving equipment to be able to decode the incoming binary pulse trains, it is only necessary for it to be able to determine whether or not a pulse is present. The process of encoding and quantitizing are reversed: a replica of the PAM signal is first generated, then this is fed in to a low pass filter to reconstruct the original analog signal.

Part C
Switching and Signalling

14 Step-by-Step Telephone Exchanges

Telephony was invented in the 1870s; all the early exchanges used human operators to establish and supervise calls. As networks grew it became uneconomic to continue to use people to set up telephone calls. It has indeed been calculated that to carry today's telephone traffic using 19th century practices would need more than half the total population of all major cities to be employed as telephone operators!

For automatic operation the first requirement is a way of indicating to the exchange the telephone number of the customer to whom you wish to speak. The rotary dial with ten finger holes is now nearly a century old in basic concept but is still in wide use. Contacts within the dial make and break an electrical circuit which interrupts current flowing, from a battery in the exchange, through the loop made by the line to the customer's premises and through the phone itself. If for example you dial 7, the dial breaks the circuit 7 times, with each break lasting a predetermined time, usually about 1/20th of a second. Relays in the exchange respond to these break signals.

The step-by-step principle was the first automatic system to become practicable for public telephone exchanges; the selection of a particular line is based on a one-from-ten selection process. For example, fig. 14.1 shows a simple switch that has ten contacts arranged around a semi-circular arc or bank, with a rotating contact arm or wiper that can be made to connect the inlet to any one of the ten bank contact outlets as required. The wiper is rotated by a simple electro-magnet driving a

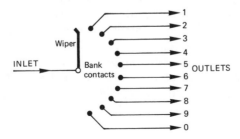

Fig. 14.1
Principle of switching
by electro-mechanical
uniselector of
one-from-ten outlets

suitable mechanism, so the arrangement is called an electro-mechanical switch.

The wiper rotates in one plane only, so this type of electro-mechanical switch is called a uniselector. Clearly the inlet can be connected to any one of the ten outlets, but the outlets are numbered from 1 to 0, which is normal practice in the step-by-step switching system.

This principle can be extended to enable the inlet to be connected to any one from 100 outlets by connecting each of the ten outlets of the first uniselector to the inlet of another uniselector, as shown in fig. 14.2. The switching of the inlet to any one of the 100 outlets (numbered 11 to 00) is done in two steps, the first digit being selected on the first uniselector, and the second digit being selected on the second uniselector.

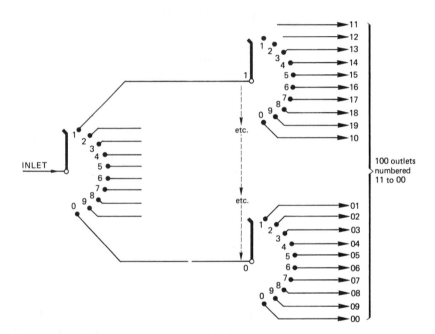

Fig. 14.2
Simple step-by-step selection of one-from-a-hundred outlets

If each of these 100 outlets is now connected to another uniselector, the inlet can then be connected to any one from 1000 outlets, numbered from 111 to 000, with the digits being selected one at a time on the three successive switching stages. This arrangement can theoretically be extended to accommodate any number of digits in a particular numbering scheme.

The same sort of numbering scheme can be provided (on a step-by-step basis) by a different type of electro-mechanical switch called a two-motion selector. The principle is illustrated simply in fig. 14.3a and b.

The bank of fixed contacts now contains 10 semi-circular arcs, each having 10 contacts, and arranged above each other. The moving contact or wiper can be connected to any one of the 100 bank contacts by first moving vertically to the appropriate level, and then rotating horizontally to a particular contact on that level. The 100 outlets are numbered from 11 to 00.

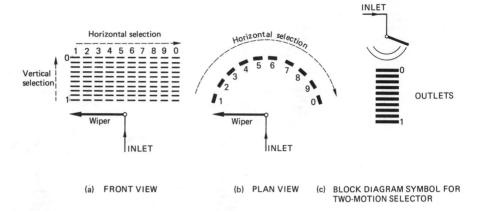

(a) FRONT VIEW   (b) PLAN VIEW   (c) BLOCK DIAGRAM SYMBOL FOR TWO-MOTION SELECTOR

Fig. 14.3
Principle of one-from-a-hundred selection by two-motion selection

The diagram symbol used to illustrate the 100-outlet 2-motion selector is shown in fig. 14.3c.

As with the uniselector arrangement, the two-motion selector system can be extended to give access to any number of outlets by adding an extra switching stage for each extra digit required in the numbering scheme. A 3-digit numbering scheme from 111 to 000 outlets is illustrated in fig. 14.4 with the one-from-a-hundred selector preceded by a one-from-ten selector.

In fig. 14.4, the first digit of the 3-digit numbering scheme raises the wiper of the first 2-motion selector to the appropriate vertical level. The selector then automatically searches for a free outlet on that level to the next selector which caters for the last two digits of the 3-digit number, as shown in fig. 14.3a.

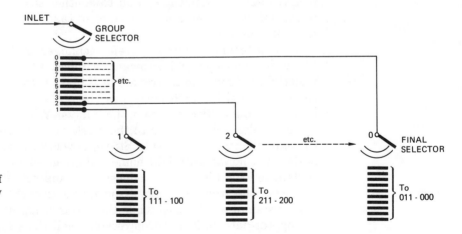

Fig. 14.4
Theoretical selection of one-from-a-thousand by two-stage step-by-step switching

In order to provide access to 10 000 lines, a further stage of group selectors is added before the final selectors. Figure 14.5 illustrates how a calling subscriber can be connected to other subscribers in an exchange having a 4-digit numbering scheme.

Theoretically, a 4-digit numbering scheme can accommodate 10 000 subscribers, but it is necessary also to provide junctions to other

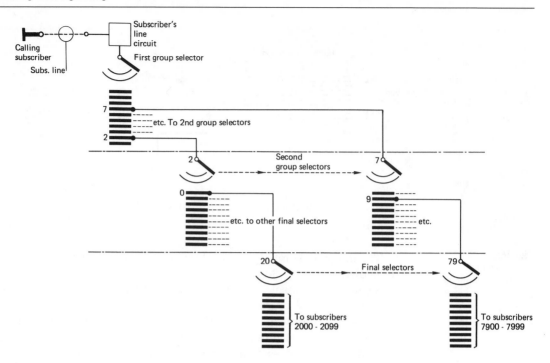

Fig. 14.5

Simple trunking diagram of 4-digit step-by-step automatic exchange

exchanges, lines to the operator and other enquiry services, and so on. This means that the capacity of an exchange of this type would in practice be about 6000 subscribers instead of the theoretical 10 000.

One of the basic features of step-by-step exchanges is that each selector or switch is controlled by a group of electromagnetic relays which is in effect a small brain, just sufficient to act on the digit it receives and to route the call on to another selector which acts on the next digit, and so on. Each digit dialled takes the caller one step nearer the called number.

In some countries step-by-step exchanges and selectors are called Strowger exchanges and Strowger switches. Almon B. Strowger was an undertaker (American mortician) in Kansas City, Missouri. The wife of the rival mortician in Kansas was the operator in the local telephone exchange (manually operated in those days). Whenever an anguished call came in "please connect me to the mortician" the rival got the call and the funeral business. Almon B. Strowger was so incensed by the injustice of this that he went straight home and invented an automatic switching system to keep himself in business. This unlikely story really is true.

Most of the group of relays associated with each selector in a step-by-step system are only used while the call is being set up, so, as soon as the appropriate digit has been received and the selector stepped to the particular number dialled, most of these relays are idle. In a main exchange there are likely to be several thousand of these complex selectors; many of them will only be brought into use during busy periods (say during morning peak traffic) and even the busiest (or "first choice") selectors only use all their "brains" for a second or two every few minutes. It follows therefore that a large amount of expensive equipment in such step-by-step exchanges is idle for most of the time.

Step-by-step selectors are robust and the principles are easy to follow, so fault finding is usually fairly straightforward and faults can normally be speedily rectified. The selectors do however sometimes introduce an unacceptable level of noise into conversations; they "shudder" quite violently while calls are being set up and while they are releasing at the end of each call. These movements affect electrical resistance at contact points and so produce noise in circuits.

Selectors have a great many moving parts which means that regular lubrication, with cleaning of relay contacts and readjustment of switches from time to time, is absolutely essential if good service is to be maintained. The high cost of maintenance is one of the main reasons for the fact that electro-mechanical step-by-step exchanges are, in most countries, now being replaced by exchanges which use low-maintenance-cost electronic techniques.

15 Reed Relay and Crossbar Exchanges

A reed relay is a device based on the fact that an electric current passing through a coil of wire produces an electro-magnet, with the ends of the coil having opposite magnetic polarities, as in fig. 15.1.

If now two thin strips of material that can be magnetized are placed inside the coil, the strips will become magnetized when the current is flowing in the coil. If the two strips are placed so that one end of each overlaps, they will have opposite magnetic polarities and so will attract each other, as shown in fig. 15.2.

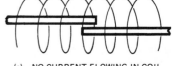

(a) NO CURRENT FLOWING IN COIL, STRIPS SEPARATED

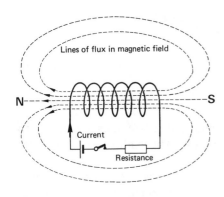

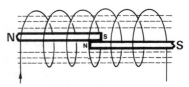

(b) CURRENT FLOWING IN COIL, STRIPS ARE MAGNETIZED AND ATTRACT EACH OTHER TO FORM AN ELECTRICAL CONTACT

Fig. 15.1
Coil of wire as a simple electro-magnet

Fig. 15.2
Principle of operation of a reed relay

These two strips can be used to form a switch in another electrical circuit. The two strips are placed inside a glass envelope containing an inert gas, and the overlapping portions are coated with gold to give a good reliable electrical contact. The whole assembly contained by the glass envelope is called a reed insert, since it is placed inside the electro-magnet coil.

A typical reed relay has four of these reed inserts placed inside the electro-magnetic coil, each of which can be used to switch a separate electrical circuit. Coils are usually then arranged in a matrix formation so that the contacts in any particular reed relay in the matrix may be operated under the control of pulses of current through winding coils.

Selectors in crossbar exchanges have horizontal and vertical bars operated by electromagnetic relay coils, so that, with a crossbar switch also, the contacts at a particular point in a matrix may be operated under the control of these relays.

Crossbar switches and reed relays are both used in telephone exchanges. The basic concept is however quite different from that of step-by-step exchanges.

Instead of each switch or selector having its own little distributed "brain", there is a central "brain" which controls all switches (see fig. 15.3). This central "brain" or register/marker is rather like a computer; it registers the number dialled, it checks that the calling number is permitted to make the call, and tests to see if the called number is engaged. Exchanges using this centralized control function are called common control exchanges. If the called number is free, this common control equipment chooses a path through the exchange to join together the calling line to the called line, and issues instructions to all the crossbar switches or reed relays concerned to operate in such a way that the two lines are connected together. All this happens very rapidly—calls do not have to be switched through the exchange one digit at a time in step with the subscriber's dial.

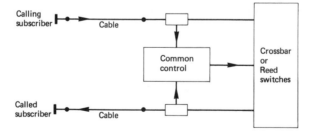

Fig. 15.3
Principle of reed relay and crossbar exchange

This increased speed of operation means that it becomes attractive to replace ordinary rotary dials by push buttons; these allow calls to be set up very rapidly indeed.

As there is less mechanical complexity in a crossbar or reed relay system, it is more reliable than step-by-step. Also the path through the exchange is such that crossbar switches and reeds introduce less noise into a telephone conversation than step-by-step switches.

Crossbar switches and reed relays have no moving parts demanding regular maintenance and adjustment so very little routine maintenance is needed and fewer exchange men are required. It is this reduction of maintenance staff which usually results in large common-control exchanges being more economic than similar size Strowger step-by-step exchanges. For small-capacity exchanges the extra cost of common control equipment may be too great to be offset by maintenance savings. Increasingly, therefore, such exchanges are provided as remote concentrator switches controlled by a larger parent exchange.

Both crossbar and reed relay switching depend on the operation of a switching matrix, the principle of which can be explained by considering the circuits which are to be connected together as being arranged at right angles to each other in horizontal and vertical lines. These lines represent inlets and outlets of the switch. This idea is illustrated in fig. 15.4.

The intersections between horizontal and vertical lines are called crosspoints. At each crosspoint some form of switch contact is needed to complete the connection between horizontal and vertical lines, as shown in fig. 15.5. Any of the 4 inlets can be connected to any of the 4 outlets by closing the appropriate switch contacts. For example,

a) Inlet 1 can be connected to outlet 2 by closing contact B.
b) Inlet 4 can be connected to outlet 3 by closing contact R.

Considering figs 15.4 and 15.5 again, it can be seen that with 4 inlets and 4 outlets there are 16 crosspoints.

Obviously, the number of crosspoints in any matrix switch can be calculated by multiplying the number of inlets by the number of outlets. This is further illustrated in fig. 15.6. If there are n inlets and m outlets, then the number of crosspoints is $(n \times m)$.

a) If n is larger than m, that is if there are more inlets than outlets, then not all the inlets can be connected to a different outlet. When all the outlets have been taken, there will be some inlets still not in use.
b) If m is larger than n, that is there are more outlets than inlets, then,

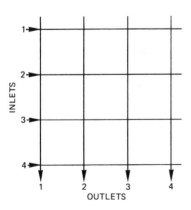

Fig. 15.4
Simple 4 × 4 switching matrix

Fig. 15.5
Principles of switching by a 4 × 4 matrix switch

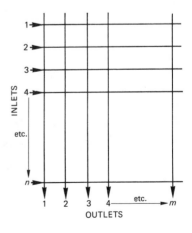

Fig. 15.6
Number of crosspoints in a matrix switch

when all inlets are each connected to an outlet, there will be some outlets still not in use.

So, the maximum number of simultaneous connections that can be carried by a matrix switch is given by whichever of the number of inlets or outlets is smaller. For example, if there are 10 inlets and 5 outlets, then the maximum number of simultaneous connections possible is 5, as illustrated in fig. 15.7.

Inlet 1 is connected to outlet 1
Inlet 2 is connected to outlet 2
Inlet 3 is connected to outlet 3
Inlet 4 is connected to outlet 4
Inlet 5 is connected to outlet 5
Inlet 6, 7, 8, 9, 10 cannot be used.

The efficiency in use of crosspoints may be calculated as

$$\frac{\text{Maximum number of crosspoints that can be used simultaneously}}{\text{Total number of crosspoints in one matrix}} \times 100\%$$

For a large matrix this efficiency is necessarily very low, e.g. a 15×15 matrix with 225 crosspoints is only able to use 15 of these at any one time, giving only 6.7% efficiency.

Efficiency can be improved by using smaller matrix switches, linked together; most crossbar and reed relay exchanges are designed on this basis, with a series of interconnected switches. This can, in some circumstances, lead to link congestion or internal blocking. Careful design of the exchange is needed in order to maximize its traffic handling capacity while minimizing equipment quantities (and therefore costs).

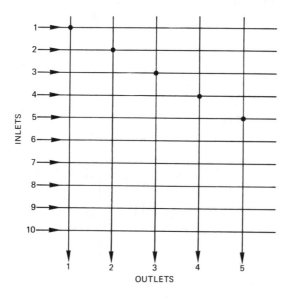

Fig. 15.7
Simple 10×5 matrix switch

16 Stored Program Control

Control in telephone exchanges developed from individual control of each switch in step-by-step exchanges to the use of a small number of complex centralized units in so-called "common control" exchanges, which were mainly crossbar or reed. Common control units were originally completely electro-mechanical, using the same basic types of relay which had been used for many years in earlier exchanges. Component improvements, in particular the development of cheaper memories, rapidly led to the introduction of new designs of these control units.

The first electronic common control units made use of thermionic valves in place of "traditional" electro-mechanical relays. These units demonstrated principles but were never considered to be really serious contenders; electronic common control had to wait for the popularization of the transistor and the printed circuit board before it became truly economic.

By the late 1960s computers were being developed very rapidly. It seemed to many of the brightest computer designers that they could readily use their computers as efficient and fully centralized common control units for telephone exchanges.

There were however a few snags. A general-purpose processor designed to perform efficiently in a business environment is not necessarily capable of performing satisfactorily all the duties required of a central processor operating in real time in a telephone exchange. The table below summarizes the main differences in design philosophies.

	General-purpose computer	*Telephone exchange central processor*
Complete stoppage	Inconvenient	Unacceptable
Faulty output	Unacceptable	Inconvenient

These inescapable factors led to the development by some telecommunications firms of wired logic units able to perform most of the required functions, while some firms developed their own computers, especially designed to perform telephone exchange control duties. Some manufacturers have concentrated on improving and developing efficient and powerful central processors alone, but in the last few years technological developments in the design of multi-function components seem to indicate a long-term trend in the telephone exchange field towards distributed control by microprocessors which refer to their parent main processors or to other micros for special functions or for facilities such as abbreviated dialling, which require access to large memories. Good signalling systems between these distributed processors could well, from a service point of view, prove just as important as the actual switching practice followed.

The use of computers to control telephone exchange switching is called Stored Program Control or SPC. This has been defined as the control of an automatic switching arrangement in which call processing is determined by a program stored in an alterable memory.

Stored program control techniques have been used for crossbar exchanges and reed relay exchanges. During the 1980s suitable solid state devices have however become available at competitive prices. Many major manufacturers now produce exchanges using completely solid state switching devices; there are no moving parts, nothing ever requires readjustment.

Until very recently the signals passing through all types of telephone exchange were analog in nature, i.e. they were continuously variable in amplitude or frequency in response to changes of sound pressure impressed on the microphone at the speaking end of a circuit. All first-generation SPC exchanges were necessarily therefore exchanges which established physical, metallic, space-division switched paths for analog signals although controlled by digital computers.

Now that the world is moving away from analog transmission and space-division switching to time-division and digital techniques, the same types of processors used to control analog SPC exchanges can still be used to control new-generation time-division switching. The increased use of digital techniques has indeed led to the increased use of computers in telephone exchanges. Not only are microprocessors now widely used in many of the subsystems from which exchanges are built up but general purpose computers are increasingly used for network management, billing and administrative tasks directly associated with each exchange.

It is also of great potential importance that business computers are now beginning to be built which provide the same standards of reliability available at one time only in special telecommunications processors. New-generation business computers can therefore be used to control telephone exchanges.

17 Memories and Stores

In most digital systems and in all computer-type applications, elements capable of storing binary-coded information are required. For example, the arithmetic/logic circuits of a computer must be able to hold results during the processing of a problem, and memories in a computer controlled telephone exchange must hold a tremendous amount of information, for example:

a) Details of every customer's line, and the equipment installed in every customer's premises.

b) Whether or not each line is authorized to initiate long-distance or international calls or is perhaps restricted to receiving calls only.

c) The digits which customers dial to call distant exchanges, and the way calls to all such exchanges have to be routed.

d) Alternative ways of routing calls to particular destinations whenever the principal routes are congested.

e) The status of every cross-point and switch in the exchange so that calls may be routed through the exchange avoiding congested or faulty sections.

f) Details such as time of day, number dialled, duration of connection for every call so that bills may be prepared.

It is therefore of fundamental importance to provide logic elements which have a memory and are capable of storing binary data for a given time so that this may be available for subsequent use when needed.

A computer usually requires different kinds of memory stores:

a) A main store: this contains all the information which must be immediately available to the processor.

b) A larger capacity backing or back-up store: this contains most of the data and programmes and is used to hold information that need not be immediately available to the processors.

c) Several registers: these provide short-term storage of information such as interim results of individual stages in complex calculations, or data read out from a main or backing store so as to be ready to be used by a processor.

It must be possible both to write information into a memory and to read information out of the memory. To make this possible a memory consists of a large number of locations—often arranged in a matrix—in each of which a small amount of data can be stored. Each location has a unique address so that it can be accessed from outside the memory. The access time of a memory is the time that is needed to read one word out of the memory, or to write one word into the memory.

The main stores in early computers were usually made up of a matrix of ferrite cores; these are small rings of ferrite through which wires are threaded which both control the magnetization of each core and read out the direction of such magnetization. Main stores very rarely use ferrite cores nowadays because semiconductor stores have become cheaper and are far more compact—but the name has remained in use, so a main store is often called a core store even though it uses no cores.

Magnetic tape and magnetic disc are widely used, especially for back-up stores where rapid and equal time access to all the locations in the store is not so important as the ability to hold information irrespective of the failure of power supplies. Both these magnetic stores work on the same principles as the domestic tape recorder.

The magnetic bubble is beginning to come into use in special applications such as telephone exchanges where an extremely large storage capacity, rapid access, and no moving parts to require maintenance are of paramount importance. Bubble memories are at present expensive devices; their manufacture demands extremely high precision.

Solid state semiconductor memories are however the types of memory which are of most immediate concern to us, they are very widely used in electronic equipment of all types.

Basic semiconductor memory elements are bistable, and are often called flip-flops or latches. When used to store a logic 1 bit, the input "flips" the memory element so that the output takes and holds the logic 1 state. It will hold this state indefinitely or until it is reset or "flopped" back to its other stable state, the logic 0 state, by a subsequent input. The several varieties of flip-flop are made in integrated circuit form, and many IC chips contain a large number of circuits made up of combinational or logic gates interconnected with memory elements.

Most systems can operate with both volatile memory (memory which forgets or loses its contents when the power supply is removed) and non-volatile memory (which retains its contents even when power is removed). In general, volatile memory is used as workspace for storing temporary data and the results of calculations during program execution, whilst non-volatile memory is used to store programs, permanent records, etc.

There are two basic types of semiconductor memory device:

1) Those into which information has been written during manufacture; this information can be read out by the user but it cannot be changed. These are called Read-only Memories or ROMs.

2) Those into which information can be written by the user, changed by the user if necessary, and read out by the user when wanted. These are called Random-access Memories or RAMS. This is not really a very good descriptive name for these devices; it is perhaps possible that, if their original designers had called them Write And Read memories (which they are), they might, as WAR Memories, never have become socially or economically acceptable in some quarters.

A ROM, then, is a memory which is permanently programmed and can only be read, i.e. program instructions, data, etc., can only be copied from ROM memory because each ROM has its information "burnt" into its store during manufacture. Typical applications are

a) Monitor programs or operating systems programs, which are the programs that control the operation of a microcomputer system and allow the user to run application programs, to input and output data, to examine and modify memory, etc.

b) Dedicated programs, each one specially devised to fulfil one particular function, such as a control program or a complex calculation routine.

c) Commercial recordings of music and video for home entertainment, using Compact Disc (CD) technology. CDs are not electronic or magnetic, they are manufactured using high-power lasers which burn tiny pits (only about 5 microns in diameter) on a tight spiral path on a hard disc, which is then covered with a transparent layer of protective plastic. The user's CD player utilizes a low-power laser which in effect follows the same spiral path as the disc spins. When it is over a pit, the pulsed laser's light is diffused and not reflected back; where there is no pit, the laser's pulse is reflected straight back. This ensures recognition of the binary coded figures burnt into the disc, representing quantized sample levels in this high-fidelity form of pulse code modulation, PCM.

ROMs are usually programmed in the factory and, since it is clearly not possible to maintain power supplies to them while they are being transported for assembly into a system, they have to be non-volatile; information stored in a ROM is not lost when power supplies are switched off.

A RAM is designed to have information both written in to it and read out of it. Random access means that each stored word can be accessed or retrieved in a given amount of time. The memory locations storing the words making up a program instruction, data, etc., can be accessed in random order in the same amount of time, independent of the position of the word in the store. In this sense, ROMs are also random access devices but programs stored on magnetic tape must be accessed sequentially and are therefore not random access. RAMS are used for storing application programs during use, fed in for example from a tape, disc or printer. They are also used for storing intermediate results during a program execution. Information stored in RAM can readily be modified. RAM memory is usually volatile and lost when power supplies are switched off; if a back-up power supply maintains power in the event of mains failure, the information in store will of course not be lost.

The ROM has been described in its basic form. There are, however, a number of important versions of it, three of which are now briefly described.

A programmable read only memory or PROM is a special type of ROM designed to be programmed by the user to suit a specific application for a device. It uses a column-and-row memory matrix with all of the intersections linked by fusible diodes (transistors). It is programmed by addressing the particular locations in memory that are to store logic 1 and passing a sufficiently large current through the transistor at that intersection to blow the fuse, so that there is no longer a link between column and row of the matrix at that particular point.

An Erasable PROM or EPROM is similar to an ordinary PROM except that the chosen matrix intersections are not permanently made logic state 1 by blowing internal fuses. This logic state is written in by the storage of an electrical charge at the point. If a program on an EPROM is to be changed, the EPROM is exposed to ultraviolet radiation directed through a small window in the package containing the chip. This removes the stored charge at every location in memory so the chip has to be completely reprogrammed.

In an Electrically-alterable PROM or EAPROM, as with the EPROM, programming a memory cell to store logic state 1 is accomplished by charging that cell. With an EAPROM, erasure procedure is carried out by applying a reverse polarity voltage to a cell to remove its charge; this erasure is done on individual cells in the matrix, it is not a complete chip erasure as it is with an EPROM.

18 Signalling

In a telephony context, signalling means the passing of information and instructions from one point to another relevant to the setting up or supervision of a telephone call.

To initiate a call a telephone subscriber lifts the handset off its rest—in American English, "goes off hook". This off-hook state is a signal to the exchange to be ready to receive the number of the called subscriber. As soon as appropriate receiving equipment has been connected to the line, the exchange signals dial tone back to the calling subscriber who then dials the wanted number. On older exchanges, this information is passed via a rotary dial by a series of makes-and-breaks of the subscriber's loop, interrupting current flow. On more modern exchanges, voice-frequency musical tones are sent to the exchange as push buttons are pressed. These tones are usually called DTMF for Dual Tone Multi-Frequency, because each time a button is pressed two tones are sent out to line simultaneously, one from a set of four high frequencies, one from a set of four low frequencies. The subscriber in due course then receives advice from the exchange about the status of the call, either a ringing signal (indicating that the wanted line is being rung), an engaged or busy tone signal (indicating that the wanted line is already busy on another call), an equipment busy tone signal (indicating congestion somewhere between the called exchange and the calling line), or some other specialized tone.

These are the signals and tones with which telephone subscribers themselves are concerned. Telephone signalling is however also concerned with the signalling of information between exchanges.

Until recently all such signalling was carried on, or directly associated with, the same speech path as was to be used for the call being established or supervised. Various terms are commonly used in connection with these speech-path-associated or channel-associated signalling systems:

a) MF—multi-frequency, i.e. using voice-frequency tones.

b) MFC—multi-frequency compelled: this type of signal continues until the distant end acknowledges receipt and compels it to stop.

c) 1VF—one voice frequency: a single tone, sometimes pulsed in step with rotary dial impulses. 2600 Hz is a common 1VF tone.

d) 2VF—two voice frequencies: two tones, sometimes used together, sometimes separately.

e) Inband—a tone actually on the voice circuit itself, audible to anyone using the circuit (and so cannot be used during conversation).

f) Outband—signals directly associated with a voice circuit but either carried on separate wires or using a different frequency, just outside the commercial speech band of 300–3400 Hz. A frequency of 3825 Hz is often used for outband signalling.

All of these signalling systems have a number of limitations:

a) Relatively slow.

b) Limited information capacity.

c) Limited capability of conveying information which is not directly call-related.

d) Inability of some systems to send detailed information back to the calling end.

e) Inability of some systems to provide sufficient information for accurate itemized call billing.

f) Systems tend to be designed for specific application conditions.

g) Systems tend to be expensive because each circuit has to be equipped independently; there are no sharing techniques and no economies of scale.

The increased use of computer-controlled (or SPC) exchanges has led to the introduction of a completely different signalling concept. Instead of signalling being carried out on or directly associated with the voice channel carrying the conversation, there is now a move towards signalling being concentrated onto fast data circuits between the processors of the SPC exchanges concerned, leaving the voice circuits purely to carry voice signals. Signalling for several hundred long-distance circuits can be carried by a single fast data system, and substantial economies result.

A signalling system of this type has now been standardized by the body responsible for drawing up specifications for international use; this is called CCITT Signalling System No. 7. (CCITT means the International Consultative Committee for Telephony and Telegraphy.) No. 7 signalling has not only been designed to control the setting-up and supervision of telephone calls but of non-voice services also, such as word processors, teletext machines, etc.

a) Signalling is completely separate from switching and speech transmission, and thus may evolve without the constraints normally associated with such factors.

b) Significantly faster than voice-band signalling.

c) Potential for a large number of signals.

d) Freedom to handle signals during speech.

e) Flexibility to change or add signals.

f) Potential for services such as network management, network maintenance, centralized call accounting.

g) Particularly economic for large speech circuit groups.

h) Economic also for smaller speech circuit groups due to the quasi-associated and disassociated signalling capabilities (see fig. 18.1).

i) Systems have been standardized for international use.

j) Can be used to control the setting-up and supervision of non-voice services and so will be important for ISDN (see section 26).

A common channel signalling system providing the signalling for many speech circuits must have a much greater dependability than channel-associated signalling since random errors could disturb a large number of speech circuits. For this reason, provision must be made to detect and correct errors. Additionally, automatic re-routing of signalling traffic to a good back-up facility must occur in a situation of excessive error rate or on failure of a signalling link.

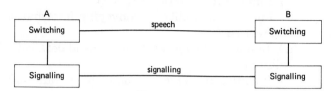

(a)   *Common-channel signalling between A and B on an associated basis*

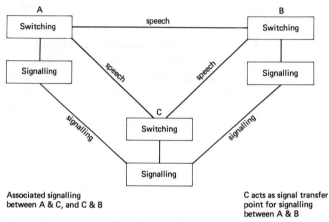

Associated signalling
between A & C, and C & B

C acts as signal transfer
point for signalling
between A & B

(b)   *Common-channel signalling between A and B on a quasi-associated basis*

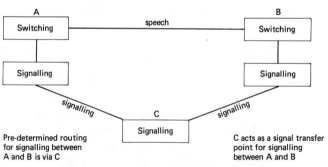

Pre-determined routing
for signalling between
A and B is via C

C acts as a signal transfer
point for signalling
between A and B

(c)   *Common-channel signalling between A and B on a dissociated basis*
      i.e. signalling can be routed and switched quite independently
      from the speech circuits being controlled

Fig. 18.1
Common channel
signalling (such as
CCITT No. 7)

19 Digital Exchanges

It has become conventional to divide telephone exchanges into two main categories:

1) *Space Division (or Analog)*—in which direct physical paths are established right through the exchange from one subscriber's line to another. Connections may be by metallic contacts (electro-mechanical step-by-step Strowger switches, rotary switches, crossbar switches or reed relays) or by solid-state analog devices.

2) *Time Division (or Digital)*—in which some or all of the switching stages in the exchange operate by shifting signals in time. Basically a connection is made between incoming and outgoing channels by transferring each PCM word (see section 13) from the time-slot of the incoming channel to that of the outgoing channel. This time-shifting is carried out by means of stores. Information is written into an address in a store, then during every cyclic scan the information from that particular address is read out so that it occupies the required out-going time-slot.

Some purists object to the phrases "digital exchanges" and "digital switching"; these exchanges should, they say, really be called time-division. But market forces are such that the word "digital" is invariably used by manufacturers all over the world, even if some of the switching crosspoints in their exchanges are in fact pure space-division switches.

The main reasons for going over to digital are

a) Lower first cost and lower annual charges than for analog equipment. Maintenance staff savings are for example likely to be significant.

b) Space savings: digital exchanges in a fully digital environment will take up far less space than analog exchanges—in some circumstances less than 1/10th of the floor area will be needed.

c) Transmission improvement: the change from FDM to digital TDM transmission systems combined with the change from 2-wire space-division exchanges to digital exchanges (which are in effect all 4-wire devices) enables losses to be reduced significantly without having to invest heavily in new cable plant in local distribution networks.

d) The relative ease by which digital switching equipment can evolve to provide the many new services which customers are beginning to demand.

Most digital exchanges are built up from subsystems; the same subsystems can be put together to provide a variety of exchanges, for use at different points in the network (see figs 19.1 and 19.2).

The simplification of function which follows the final elimination of FDM and analog trunks and junctions will be apparent. When this

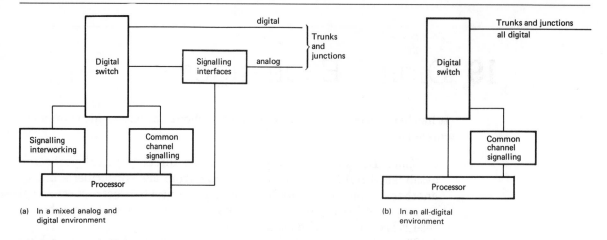

(a) In a mixed analog and
digital environment

(b) In an all-digital
environment

Fig. 19.1
Digital trunk exchange

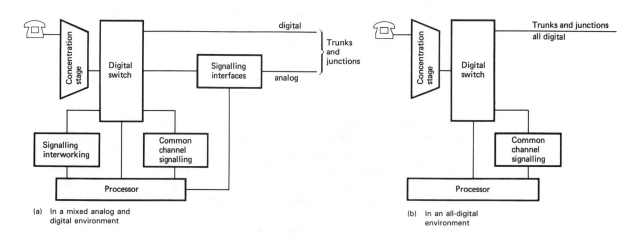

(a) In a mixed analog and
digital environment

(b) In an all-digital
environment

Fig. 19.2
Digital local exchange

stage is reached, exchanges will all take up a fraction of the floor area
now occupied, with great consequent savings.

The function of the concentration stage is to interconnect subscriber
lines with the main switching subsystem in the exchange, the digital or
group switching stage. Out of every 1000 lines on an exchange, the probability is that not more than about 100 will be making calls at any one
instant of time, even during busy periods. It would therefore be a waste
of money to provide so much switching equipment that all the subscribers on the exchange would be able to talk at the same time. The concentration stage has to be carefully designed to provide the grade of
service required by the administration, with the greatest possible
economy. Business lines are used much more than residential lines, so it
is not possible to concentrate business lines to the same extent as low-
calling-rate residential lines. (Junctions to other exchanges are not concentrated at all; every trunk and junction circuit always has direct
access to the group switch.)

This principle of concentration of subscriber's lines, followed by switching, followed by a reverse path through the concentrator out to the

called subscriber, is not peculiar to digital exchanges. Step-by-step, crossbar and reed relay exchanges also use the same type of switching philosophy—note here however that, when a concentration stage serves the called subscriber, it is sometimes called the expansion stage, and shown separately even though it is sometimes the same piece of equipment as the concentration stage. In digital exchanges it is usual to give special consideration to the concentration stage because most types of digital exchange use PCM systems between concentrators and their digital group switching stages. In some makes of exchange the matrix in the concentration stage is completely different in design and technology from that in the group switching stage. Some makes of digital exchange do indeed use analog space-division switches in their concentration stages, to reduce total costs. In electromechanical (including common control) systems, the same general type of switch is used in all stages of switching: those used for concentration have more inlets than outlets; those used for switching usually have the same number of inlets as outlets; those used for expansion stages have more outlets than inlets. (Since the same technology is used for all stages it is not usual to consider concentration stages as being anything other than part of the switching matrix in these earlier types of exchange.)

The interface with the subscriber's line is at the present time the most costly part of all digital exchanges, largely because a 10 000 line exchange has to have 10 000 of these units, each with all the features needed to interwork with various types of subscribers' lines. It is customary to describe these features as the "Borscht" functions, based on the initials of the key words:

> Battery feed to line—there is normally no active power source at a subscriber's telephone; all the power (needed to drive the microphone and key pad) is fed out from the exchange along the subscriber's line.
>
> Overvoltage protection—solid state devices are very sensitive to high voltages, and rapid-action protective devices have to be provided in each line circuit so that if lightning does happen to strike an external line the exchange will not be put out of action.
>
> Ringing current injection and ring trip detection—the bell at the called subscriber's telephone has to be rung; this means that quite a high voltage a.c. signal (sometimes about 70 volts) has to be connected to the line, to ring the bell; and as soon as the handset has been picked up off its cradle (gone "off hook") the ringing must be tripped and disconnected.
>
> Supervision of the line—equipment has to be provided which continually monitors the line so that as soon as it goes "off hook" (and a continuous d.c. path provided through the instrument) the exchange connection is activated and dial tone sent out to the caller. Dial pulses represent breaks in the continuous d.c. loop; these have to be detected and counted so that the exchange knows what number is required. When the caller finally clears down, by going back "on hook", the exchange must break down the established call and note the time at which this has been done, for charging purposes.
>
> Codec (short for encoder plus decoder)—this turns the analog signal

received from the telephone instrument into a digital signal ready to be multiplexed with others into a PCM system. Incoming signals are similarly decoded from digital to analog before being sent out to the subscriber's instrument. Some digital exchanges have one codec per line; some share codecs between several lines; some use even fewer codecs by placing them between their concentration stages and the links going to their group switching stages.

Hybrid for 2-wire to 4-wire conversion. The line to an ordinary telephone subscriber uses one pair of wires, for both directions of conversation; this is called a 2-wire circuit. The circuits inside a digital exchange use two electrically separate paths, one for each direction of transmission; this is called a 4-wire circuit. To join together a 2-wire and a 4-wire circuit, a special device called a hybrid is used; this allows speech from the 2-wire line to enter the 4-wire transmit path, and speech from the 4-wire receive path to pass to the 2-wire line, but it blocks speech incoming on the 4-wire receive path from going out again on the 4-wire transmit path of the same circuit.

Testing of both line and equipment—it is necessary to be able to test the subscriber's line electrically so that faults may be located and cleared.

Any form of time-division switching is necessarily a one-way function; to provide a single bi-directional speech circuit through an exchange, two channels therefore have to be switched through, one for each direction of transmission. Figure 19.3 shows how this is done in some exchanges.

The use of common channel signalling systems such as CCITT No. 7 will enable a single exchange to be used to switch all types of digital information services; it will not be necessary to have separate exchanges for record services such as telex. (See section 42.)

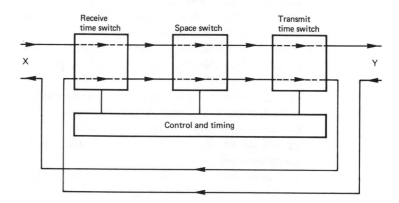

Fig. 19.3
Time-space-time switching

Part D
Telecommunication Systems Characteristics

20 Local Distribution Networks

Since we are all convinced that good telecommunications services not only pay for themselves and earn handsome profits, and are essential if national economies are to prosper, why is it that there are huge waiting lists for telecommunications services in many areas?

The most common reason of all for the existence of large waiting lists for telephones is a low level of investment in plant, particularly in the cables of outside distribution plant.

There is a certain amount of glamour attached to modern digital exchanges: good salesmen and hungry politicians can sometimes get together with the result that new buildings and new switching centres are very rarely found to be the problem areas. It is unseen and unthought of underground cables that are most often found to be the bits lacking.

This is especially likely to be the problem where telecoms systems are government-owned, and financed on an annual budget basis, making long-term projects with low initial revenue very difficult to get approved.

Whenever investment capital is in short supply and revenue growth is high priority it is natural for such funds as do get made available to be used in the large cities where new lines can be provided at lower cost than in rural areas; these latter usually need substantial initial investment and yield relatively low revenues, compared with works in urban concentrations.

It is clearly socially desirable for the rural areas in a country to be provided with services which make life acceptable for today's young people, but social engineering on a significant scale requires a political commitment. Those ultimately responsible have to be able to resist providing more funds for cities and to divert a reasonable proportion to rural development areas.

This means that the study merely of total figures in a country's waiting lists is not of itself of great value. Detailed study must be made to find out where the waiters are located—and if there are very few waiters in rural areas, is it because the tariff is so loaded against these areas that rural workers cannot afford phones?

The connection between a subscriber and the local telephone exchange consists of a pair of wires in a telephone cable. Modern networks will increasingly use fibre local distribution systems (see section 24). This use of fibre makes little or no change to the principles of the local distribution network discussed in this section. The use of fibre in the local network will demand rather than make it advisable to have concentrators located, remote from the local exchange, within the local network. Since a large telephone exchange may have 10 000 or more subscribers, the local line network can be quite complicated, particularly because provision must be made for fluctuating demand. The local line network is provided on the basis of forecasts made of the future demand for telephone service, the object being to provide service on demand and as economically as possible. Since the demand fluctuates considerably there is the problem of forecasting requirements and deciding how much plant should be provided initially and how much at future dates. No matter how carefully the forecasting is carried out, some errors always occur and allowance for this must be made in the planning and provision of cable, i.e. the local line network must be flexible. A network must be laid out so that the situation should not arise where potential subscribers cannot be given service in some parts of the exchange area while in other parts spare cable pairs remain.

The modern way of laying out a local line network is shown in fig. 20.1. Each subscriber's telephone is connected to a distribution point, such as a terminal block on a pole or a wall. The distribution points are connected by small distribution or secondary cables to cabinets. Primary or main cables then connect these cabinets to the telephone exchange.

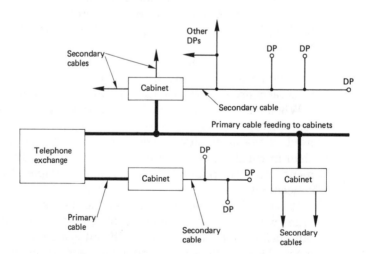

Fig. 20.1
Layout of a telephone exchange area

It is usual to provide secondary cables on the basis of an expected life of about 15 years and the much larger primary cables for only about 5 years. This does not mean that these cables are expected to be scrapped after these periods; they are expected by then to be fully utilized and to require supplementing by additional cables. Cabinets are sometimes called "flexibility points": if demand is much heavier than forecast in one part of the area served by the cabinet, and less in another part, cable pairs may be connected through at the cabinet to the faster developing area.

External plant usually represents the largest single component of the capital assets of a national telephone system so it is important that full consideration be given to this subject. All too often, however, local cabling is given only low management level attention despite its importance to subscribers and to the financial well-being of the administration. Capital expenditure has to be minimized while providing sound engineering combined with flexibility.

From the outset, line planning specialists must liaise with transmission and switching planners to designate the fundamental local network parameters in accordance with CCITT recommendations, keeping in mind the economic advantages of apportioning the largest amount possible of reference equivalent to the local distribution network. On the other hand the finer the gauge cable that can safely be used in the network, the lower the costs will be.

Using this information together with the forecast demand, the practical wire centres for each exchange location can be determined—using a computer programme if necessary. Economic studies should be carried out if the proposed site for a new exchange proves to be a long way away from the calculated wire centre. Types of construction to be used, sizes of cables and conductor gauges must then be chosen, taking into account the geological and topographical conditions in both urban and rural areas.

It is usually economically desirable to use underground ducted cable from exchange to flexibility points (cabinets) and then either go underground or overhead to distribution points depending on circumstances (e.g. in some countries it is possible to arrange for joint use of poles with the local electricity supply authority). A ducted system, although initially more costly than a direct buried system, provides much greater flexibility for the future installation of additional cables, and facilitates cable repairs or replacement.

At the present time, polyethylene-insulated conductors and polyethylene-sheathed cable incorporating an aluminium screen and water vapour barrier is the most economical type of cable to be used in ducts. Main or primary cables from exchanges to cabinets should wherever possible be airspaced with protection provided by continuous-flow dry-air pressurization. Secondary or distribution cables from cabinets to distribution points (and smaller main cables) should be jelly-filled, not pressurized. Aerial cables should be of similar design but with an in-built self-supporting catenary wire in the figure-of-eight mode.

Many modern telephone switching systems incorporate concentration stages which can be either co-located (i.e. in the same building as the rest of the exchange) or remote (i.e. many kilometres away, fed by PCM systems back to the main part of the exchange). The use of remote concentrators to serve telephone subscribers in small towns means that cable pairs to such subscribers will not in future need to be of the heavy gauge which was in the past needed to feed them all the way back to the nearest city; comparatively fine gauge (and therefore cheaper) cables can

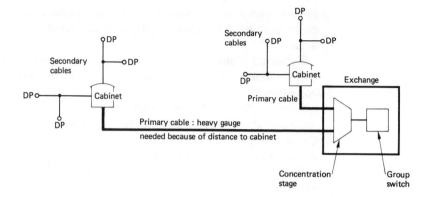

Fig. 20.2
Traditional-type cable
distribution

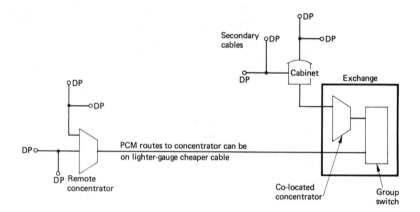

Fig. 20.3
Use of a remote
concentration stage

now be used for distribution to these subscribers. (See fig. 20.2 and fig. 20.3.)

The economic reasoning behind the use of remote concentrators may be summarized:

a) The transmission losses which can be accepted in the circuits between subscribers and exchange can be largely allocated to the secondary cable network (between concentrator and distribution point serving the subscriber) because the circuits between concentrator and exchange are low-loss channels in PCM systems. This means that a lighter gauge (and therefore cheaper to purchase, handle and lay) cable may be used in this extensive part of the distribution network.

b) Most concentrators are designed to provide an efficient concentration ratio of about 10 to 1, i.e. for every 1000 subscribers lines connected, about 100 PCM channels are needed in service between concentrator and exchange. 100 channels may be provided by only four 30-channel PCM systems, needing 8 cable pairs between concentrator and exchange. These pairs may be of comparatively light gauge so long as the 2 Mbit/sec PCM system may be established over them. This compares with 1000 cable pairs needed, possibly of heavier gauge, if there is no remote concentration.

c) The use of PCM systems on small cables out to a concentrator (instead of using 1000 pr, 2000 pr or 3000 pr cables for primary distribution) means that there is no need for the large underground manholes and jointing chambers which have to be constructed when very large distribution cables are used. Joints in these large cables take up a great deal of space.

d) Digging up roads, laying of nests of ducts, and reinstating road surfaces is very expensive. It is usually economically desirable to lay enough ducts at one time to last for about 20 year's growth and development. For primary distribution networks, in which most cables are so large that each one occupies a complete duct on its own, this means that most of the ducts in any newly-laid nests will be empty for many years—duct nest sizes have to be based on forecasts of future demand. The traditional type of distribution network does necessarily therefore involve a substantial "burden of spare plant", i.e. most of the newly buried plant in a network is non-revenue earning. With the use of smaller cables carrying PCM, all new duct nests can be made much smaller than they would have been—and more than one main cable can usually be pulled in to a single duct. In practical terms a single 4 inch/10 cm diameter duct can accommodate only one 2000 pair primary distribution cable so, if there is a forecast demand for 20 000 lines in an area, at least 10 ducts will have to be laid to serve it. If PCM and remote concentrators can be used, these 20 000 subscribers could be fed satisfactorily by about 2000 channels. These could be provided by only 70 30-channel PCM systems, needing only 140 cable pairs to carry them. With the special cables now available which are able to carry PCM systems on every pair without mutual interference, these 140 pairs would not even fill a single duct. Were fibre cables to be used, then a simple fibre would suffice. For reliability, at least two fibre cables would be provided, routed, if possible, in different duct routes.

It is of course not entirely a one-sided argument; remote concentrators need accommodation and power supplies, and have to be maintained. Also, if the concentrator itself fails, or the PCM links to the concentrator fail or are cut, all telecommunications services are lost to the subscribers served by the concentrator. It may be worthwhile, and still economic, to route the exchange to concentrator links via two different main cable routes and concentrator designs may include an element of redundancy. As time goes on it seems probable that digital concentrators will become more and more compact; they can already be accommodated in roadside cabinets like the cable cross-connection cabinets now in general use and hopefully it will one day become possible for them to be put into sealed canisters which can be jointed in to the cable network and installed inside ordinary underground jointing chambers and manholes.

In the past, most telecoms administrations all over the world have regarded local line planning as a function which can safely be carried out separately from other system planning duties. Now that it is becoming economically attractive for the "front end" of each exchange to be located close to the subscribers it serves, it is no longer possible to treat such planning in isolation.

21 Carrier Working: Groups and Supergroups

In section 9 it was shown how a speech signal can, by amplitude modulation, be changed in frequency from its original audio frequency (of 300–3400 Hz) up to a higher "carrier" frequency. Most of the world's long-distance telephony systems utilize 12-channel groups: twelve voice channels are all changed in frequency in this way so that a complete group of 12 channels occupies a 48 kHz bandwidth, basically from 60–108 kHz.

Figure 21.1 is a block schematic of the transmitting equipment required for channels 1 and 2 of a standard 12-channel group. The audio input signal to a channel is applied to a balanced modulator together with the carrier frequency appropriate to that channel. The input attenuator ensures that the carrier voltage is 14 dB higher than the modulating signal voltage, as required for correct operation of the modulator. The output of the modulator consists of the upper and lower sideband products of the amplitude modulation process together with a number of unwanted components.

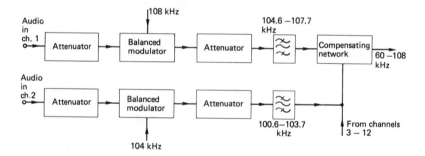

Fig. 21.1
Scematic diagram of transmitting equipment for a carrier system

Following the modulator is another attenuator whose purpose is two-fold: firstly, it ensures that the following band-pass filter is fed from a constant-impedance source—a necessary condition for optimum filter performance—and secondly, it enables the channel output level to be adjusted to the same value as that of each of the other channels. The filter selects the lower sideband component of the modulator output, suppressing all other components. To obtain the required selectivity, channel filters utilizing piezoelectric crystals are employed. The outputs of all the twelve channels are combined and fed to the output terminals of the group. The transmitted bandwidth is 60.6–107.7 kHz, or approximately 60–108 kHz.

The equipment appropriate to channels 1 and 2 at the receiving end of the 12-channel group is shown in fig. 21.2. The composite signal received from the line, occupying the band 60–108 kHz, is applied to the twelve,

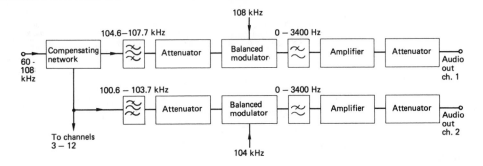

Fig. 21.2

Receiving equipment for a carrier system

paralleled, channel filters. Each filter selects the band of frequencies appropriate to its channel, 104.6–107.7 kHz for channel 1, and passes it to the channel demodulator. The attenuator between the filter and the demodulator ensures that the filter works into a load of constant impedance. The demodulator is supplied with the same carrier frequency as that suppressed in the transmitting equipment. The lower sideband output of the demodulator is the required audio-frequency band of 300–3400 Hz and is selected by the low-pass filter. The audio signal is then amplified and its level adjusted by means of the output attenuator.

The assembly of the basic 12-channel carrier group can be illustrated by means of a frequency spectrum diagram. The spectrum diagram of a single channel is given in fig. 21.3; the actual speech bandwidth provided is 300–3400 Hz but a 0–4000 Hz bandwidth must be allocated per channel to allow a 900 Hz spacing between each channel for filter selectivity to build up. Figure 21.4a shows the frequency spectrum diagram for the 12 channels forming a group; the carrier frequency of each channel is given and so are the maximum and the minimum frequencies transmitted. It can be seen that all the channels are inverted; that is, the lowest frequency in each channel corresponds to the highest frequency in its associated audio channel, and vice versa. Since all the channels are inverted, the group may be represented by a single triangle as shown by fig. 21.4b.

Fig. 21.3

Bandwidth for a commercial speech circuit

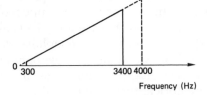

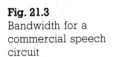

The 12-channel system can be used as a building block for the next larger assembly stage or as a system which can be transmitted to line in its own right.

Five 12-channel groups can be combined to form a 60-channel supergroup, and five supergroups make up a 300-channel mastergroup. Three mastergroups then make up a 3872 kHz bandwidth supermaster group. Alternatively, 15 supergroups may be assembled direct to form a hypergroup, sometimes called a 15 supergroup assembly.

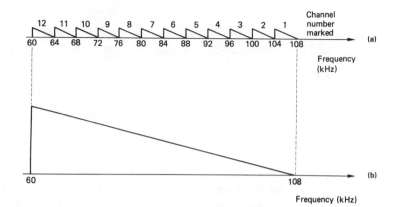

Fig. 21.4
Frequency spectrum
diagrams of a
12-channel group

22 Submarine Cables

Submarine telecommunication cables have been with us for many years; the first such submarine cable of any significant length was laid in 1850, and the first specialized cable-laying ship was launched in 1872. But early cables were all for telegraphy; the distortionless wider band needed for telephony could not economically be transmitted over long lengths of cable until electron tube and component designers had produced long-life trouble-free units, capable of operating for many years in the depths of the oceans without fault incidence.

These units (called repeaters) were designed to amplify incoming signals to offset the transmission losses incurred in the previous section of cable. It was also necessary to provide equalization to offset the way different frequencies in the wideband information signal had been attenuated by different amounts. The first submerged repeater was laid in an Irish Sea cable in 1943. They were produced in the USA in the 1950s, and repeaters suitable for insertion in deep-sea cables were soon designed in America, Britain, France and Germany.

The first long-distance submarine telephone cable was TAT-1, laid across the Atlantic from Scotland to Newfoundland in 1956. It had a capacity of 36 voice circuits. (It is no longer in use.) With the advance of technology using optical fibre and digital transmission it has become possible for wider bandwidths, meaning more circuits, to be transmitted: the latest transatlantic cable, TAT-9, brought into service in 1992, has an effective capacity of 115 000 voice circuits.

A submarine cable is sometimes under considerable tension, especially when it is being picked up from the sea bed, so great tensile strength is necessary in addition to the ability of cable and repeaters to withstand the high pressures of deep waters. Early submarine telephone cables were made up in a generally similar way to 19th century submarine telegraph cables, with one or more layers of heavy steel armour wire to provide the necessary protection (fig. 22.1). The cores of the two types of cable were of course quite different; telegraph cables usually had a heavy, well-insulated central copper conductor to carry the signal current, with the return current normally flowing through the sea itself, whereas wide-band telephone cables are of coaxial type, i.e. they have a central copper conductor, surrounded by a carefully dimensioned insulant, then copper tapes as a return-path outer conductor, then more insulant, then the outer armour wires (fig. 22.2).

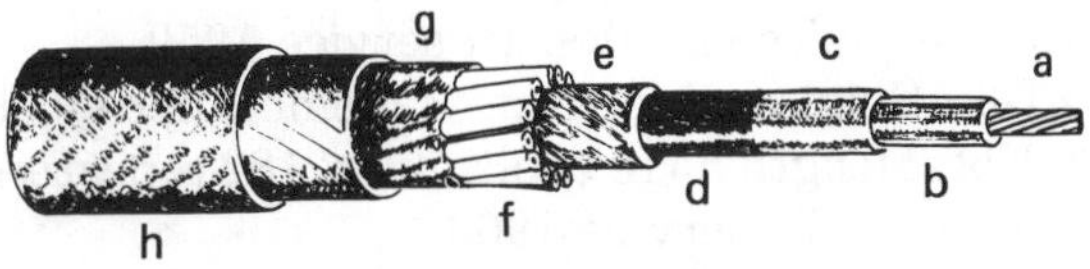

Fig. 22.1
Telegraph cable
a copper conductor
b gutta percha
  insulation
c brass tape
d bituminous wax tape
e jute
f wire armouring
g compounded tape
h jute servings with
  protective compound

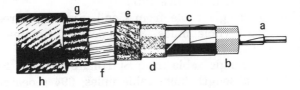

Fig. 22.2
Coaxial telephone cable
a central copper
a conductor
b polythene insulation
c return copper conductor
d impregnated fabric tape
e jute serving
f armouring wires
g impregnated fabric tape
  on each wire
h jute outer serving

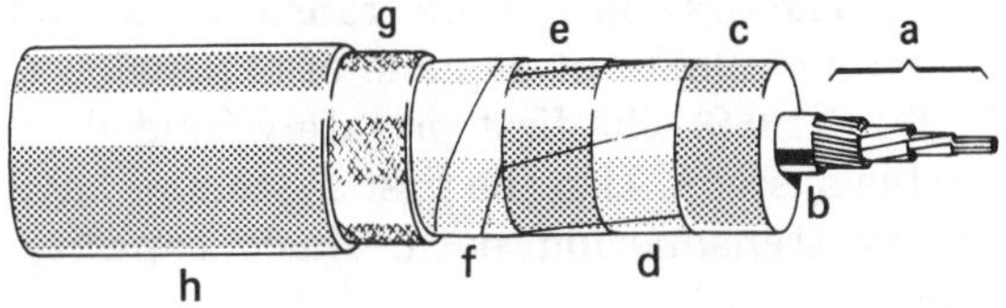

Fig. 22.3
Lightweight coaxial telephone cable
a composite high
  tensile steel stress
  member
b central copper
  conductor
c polythene insulation
d aluminium return
  conductor tape
e polythene film
  separator
f aluminium screening
  tapes with polythene
  film interleaved
g impregnated
  protective cotton tape
h polythene sheath

In the 1960s there was a breakthrough in submarine cable design for use in deep waters (e.g. deeper than about 400 fathoms): instead of cables being made up with steel wire armouring on the outside of the cable, a comparatively lightweight high tensile steel rope was put at the centre of the cable, with a circumferential strip of copper compacted on to the steel rope to act as the central conductor, with a layer of polyethylene insulant, then a layer of aluminium or copper as the return conductor, then more polyethylene, on the outside for protection (fig. 22.3). In shallow waters it is still considered safer to continue to use conventionally armoured cables, but the introduction of this lightweight design for the long deep sea sections of cross-ocean routes gave submarine cable systems a new lease of life. Cable systems are now used together with satellite radio systems. There is an economic and security need for both types of communications media to be used; most countries feel that it would be unwise to depend completely on any one system, and fibre optics now enables cable manufacturers to provide circuits even more economically than before (fig. 22.4).

The lists which follow give brief descriptions of many of the modern

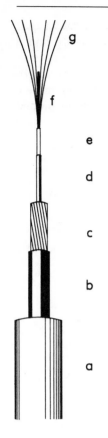

Fig. 22.4
Optic fibre submarine
cable: this provides 3
fibre pairs, each pair
able to carry up to 4000
individual 64 kbit/s
voice circuits.
Processing equipment
can increase cable
capacity to 60 000 voice
circuits per cable.
a low density
  polyethylene
b welded copper tube
c steel strands
d nylon sheath around
  core
e elastomer
f steel kingwire at
  centre of cable
g 6 fibres, in the
  elastomer cylinder

Transatlantic and Transpacific submarine telecommunications cables:

Submarine Cable Systems across the North Atlantic
TAT-1 The first transatlantic telephone cable. Twin cables from Oban, Scotland to Clarenville, Newfoundland, then single cable to Sydney Mines, Nova Scotia, Canada. Total length 4210 cable miles, 118 repeaters, nominal capacity 36 voice circuits. In service in 1956, retired in 1978.

TAT-2 Twin cables from Penmarch, France to Clarenville, Newfoundland. Total length 4418 cable miles, 114 repeaters, nominal capacity 48 voice circuits. In service 1959.

TAT-3 Single cable from Widemouth Bay, England to Tuckerton, USA. Total length 3518 cable miles, 182 repeaters, nominal capacity 138 voice circuits. In service 1963.

TAT-4 Single cable from St Hilaire de Riez, France to Tuckerton, USA. Total length 3599 cable miles, 186 repeaters, nominal capacity 138 voice circuits. In service 1965.

TAT-5 Single cable from Conil, Spain to Green Hill, USA. Total length 3461 cable miles, 361 repeaters, nominal capacity 845 voice circuits. In service 1970.

TAT-6 Single cable from St Hilaire de Riez, France to Green Hill, USA. Total length 3396 cable miles, 694 repeaters, nominal capacity 4000 voice circuits. In service 1976.

TAT-7 A 4000 circuit cable between USA and UK. In service 1983.

TAT-8 The first optic fibre transatlantic cable. Initial capacity 8000 voice circuits (1988) but capable of expansion to 40 000 circuits.

PTAT-1 (1989) The first privately funded transatlantic digital fibre optic cable system. Also the first system to have the added reliability of a standby transmission path. Basic capacity 17 000 voice circuits, maximum 115 000.

SCOT-ICE plus ICE-CAN A single cable, 24 voice circuit system installed in 1962, between Gairloch (Scotland), the Faeroes, Iceland, Greenland and Corner Brook, Newfoundland. Total length 2462 cable miles, 104 repeaters.

CANTAT-1 Single cable from Oban, Scotland to Corner Brook, Newfoundland. Total length 2073 cable miles, 90 repeaters, nominal capacity 80 voice circuits. In service 1961.

CANTAT-2 Single cable from Widemouth Bay, England to Beaver Harbour, Canada. Total length 2805 cable miles, 473 repeaters, nominal capacity 1840 voice circuits. In service 1974.

CANTAT-3 Planned for 1994.

COLUMBUS Single cable from Canary Is., Spain to Camuri, Venezuela. Total length 3239 cable miles, 503 repeaters, nominal capacity 1840 voice circuits. In service 1977.

BRACAN-1 Single cable from Canary Is., Spain to Recife, Brazil. Total length 2649 cable miles, 138 repeaters, nominal capacity 160 voice circuits. In service 1973.

Submarine Cable Systems across the Pacific

HAW-1 Twin cable from Point Arena, California, USA to Hanauma Bay, Hawaii. Total length 2210($\times$2) cable miles, 114 repeaters, nominal capacity 51 voice circuits. In service 1957.

HAW-2 Single cable from San Luis Obispo, California, USA to Makaha, Hawaii. Total length 2383 cable miles, 123 repeaters, nominal capacity 142 voice circuits. In service 1964.

HAW-3 Single cable from San Luis Obispo, California, USA to Makaha, Hawaii. Total length 2379 cable miles, 248 repeaters, nominal capacity 845 voice circuits. In service 1974.

HAW-4 In service. Basic capacity 11 500 voice circuits, maximum 57 600.

HAW-5 In service. Basic capacity 15 360 voice circuits, maximum 76 800.

TRANSPAC-1 Single cable from Makaha, Hawaii to Midway, then Wake, then Agana, Guam, then Ninomiya, Japan, with a spur from Guam to Baler, Philippines. Total length 6750 cable miles, 356 repeaters, nominal capacity 128 voice circuits. In service 1964.

TRANSPAC-2 Single cable from Makaha, Hawaii to Guam, then Chinen, Japan. Total length 4880 cable miles, 490 repeaters, nominal capacity 845 voice circuits. In service 1975.

COMPAC Single cable from Vancouver, Canada to Keawaula Bay, Hawaii, then Suva, Fiji, then Auckland, New Zealand, then Sydney, Australia. Total length 8230 cable miles, 322 repeaters, nominal capacity 80 voice circuits. In service 1963.

SEACOM Single cable from Cairns, Australia to Madang, Papua New Guinea, then Guam, then Hong Kong, then Kota Kinabalu, Sabah, Malaysia, then Singapore. Total length 7085 cable miles, 353 repeaters, nominal capacity 80 voice circuits (160 between Guam and Australia). In service 1965/67.

ANZCAN Single cable from Vancouver, Canada to Hawaii, Fiji, Norfolk Island and Sydney, Australia, with spur from Norfolk to Auckland, New Zealand. Total length 15 000 km (8000 cable miles) with over 1000 repeaters and capacity of 1380 voice circuits. In service 1984.

23 Optic Fibres

The use of a light wave as carrier, to be modulated by an information signal in the same way as these signals can modulate radio waves, was for many years considered impracticable. The reason for this was that light was emitted as a random series of energy pulses, generated largely by accelerating electrons and by electrons changing their energy levels inside atoms. Even light from what seemed to be one-colour sources like a sodium vapour lamp was found not to be a single simple sinusoidal wave but a whole series of waves, all differing in phase (and sometimes also in frequency) from each other.

The laser (light amplification by the stimulated emission of radiation) and the LED (light-emitting diode) were the new technology inventions which produced an answer to this problem. They are both devices which emit optical radiation as a direct result of applied voltages and electron movement. Both can be pulsed on and off very rapidly, and some lasers can also produce coherent light, i.e. light of a single frequency with all the waves in phase.

Although many lasers and LEDs are able to produce outputs in the visible light band, most current optic fibre telecommunications systems use signals of wavelengths 0.8 μm or 1.3 μm, both in the infra-red band. These are still called optical systems: even though the signals cannot be seen, they are transmitted in exactly the same way as visible light signals. The main reason why engineers wanted to be able to modulate coherent light was to take advantage of the tremendous bandwidths which could be carried by these very high frequencies; the figures below are typical figures, indicating the orders of magnitude involved.

	Carrier Wave		Possible bandwidth per system
	Frequency	*Wavelength*	
HF radio	3 MHz	30 m	16 kHz (4 voice channels)
Microwave radio	6 GHz	5 cm (10^{-2} m)	4 MHz (960 voice channels)
Digital microwave radio	140 Mbit/s	5 cm	64 QAM giving 40 MHz (2100 voice channels)
Optic fibre	100 000 GHz	1 μm (10^{-6} m)	Several thousand MHz (Hundreds of thousands of voice channels)

A slight complication here: when we use HF radio or microwave bearers we modulate the carrier frequencies concerned, using amplitude or frequency modulation. With light there is at present no easy way of producing beautiful smooth sine-wave bearers capable of being modulated in any of the "traditional" ways. Light is instead pulsed, either ON or OFF, with each pulse containing many hundreds or even thousands of cycles.

An optical fibre cable consists of a glass core that is completely surrounded by a glass cladding. The core performs the function of transmitting the light waves, while the purpose of the cladding is to minimize surface losses and to guide the light waves. The glass used for both the core and the cladding must be of very high purity since any impurities present will cause some scattering of light to occur. Two types of glass are commonly employed: silica-based glass (silica with some added oxide) and multi-component glass (e.g. sodium borosilicate). (Some new optic fibres do not use glass at all, but special types of plastic; these are usually cheaper to make than very pure glass but introduce greater attenuation.) Fibres now being manufactured are so free from impurities that very little energy need be lost as the signals travel along—an attenuation of less than 1 dB per kilometre is not uncommon for the latest high-purity silicons.

A major constraint with optic fibres (apart from the straightforward one of attenuation) is that, since the wavelength of light is very short, a light wave signal injected into one end of an optic fibre (sometimes called an optic wave-guide) does not merely travel straight down the middle of the core. It swings from side to side, continually being reflected or refracted back from the core/cladding surface. Clearly, if all the pulses travel at the same velocity, then a pulse going straight down the middle will reach the end just before those parts of the same pulse signal which have zigzagged along, taking a longer path. This places a restriction on the maximum possible bit rate that may be transmitted satisfactorily in the fibre.

It would have been ideal if right from the beginning we could have used fibres made with tiny diameters that were comparable with the wavelength of the optical signal being used, so that no zigzagging was able to take place, but the manufacture and jointing of such high precision fibres used to present considerable difficulty. Today's common thicker-core fibres are called multimode (because many different modes of transmission are possible). Fibres with very small diameter cores are called monomode (because only a single mode of transmission is possible). There are, therefore, three basic types of optical fibre:

1) *Stepped-index multimode* The basic construction of a stepped-index multimode optical fibre is shown in fig. 23.1*a* and its refractive index profile is shown by fig. 23.1*b*. It is clear that an abrupt change in the

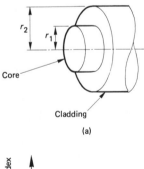

Core

Cladding

(a)

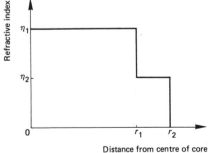

Fig. 23.1
(a) Stepped-index multimode optical fibre;
(b) refractive index profile

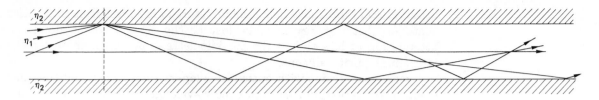

Fig. 23.2
Multimode propagation
in a stepped-index fibre

refractive index of the fibre occurs at the core/cladding boundary. The core diameter $2r_1$ is usually some 50–60 μm but in some cases may be up to about 200 μm. The diameter $2r_2$ of the cladding is standardized, whenever possible, at 125 μm.

Stepped-index multimode fibre produces large transit time dispersion (fig. 23.2), so its use is restricted to applications such as those involving comparatively low-speed data signals.

2) *Stepped-index monomode* Fig. 23.3a shows the basic construction of a stepped-index monomode optical fibre and fig. 23.3b shows its refractive index profile. Once again the change in the refractive indices of the core and the cladding is an abrupt one but now the dimensions of the core are much smaller. The diameter of the core is of the same order of magnitude as the wavelength of the light to be propagated; it is therefore in the range 1–10 μm. The cladding diameter is the standardized figure of 125 μm.

Stepped-index monomode fibres were difficult and expensive to manufacture and to join, so most of the early stepped-index optic-fibre telecommunications systems used multimode fibres. However, technological difficulties have been overcome and the use of monomode fibres is increasing throughout the world.

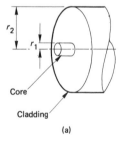

(a)

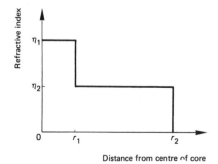

Fig. 23.3
(a) Stepped-index
   monomode optical
   fibre;
(b) refractive index
   profile

(b)

Fig. 23.4
Monomode propagation
in a stepped-index fibre

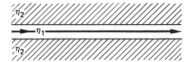

3) *Graded-index multimode* The basic construction of a graded-index multimode optical fibre is the same as that of the stepped-index multimode fibre shown in fig. 23.1a. The core diameter is also in the range 50–60 μm and the cladding diameter 125 μm. Core and cladding are mixed together in a carefully controlled manner during manufacture so that the refractive index of the inner region of the core is highest at the centre and then decreases parabolically towards the edges (fig. 23.5) to that of the cladding material. This means that light waves will be refracted back from the outer boundary of the fibre, not reflected as with stepped-index fibres (see fig. 23.6). So waves will go straight down the centre of the core, or zig-zag from side to side as they do in stepped-index fibres but in a "smoother" manner. The main difference is, however, that waves which zig-zag along in a graded-index fibre pass through regions with a lower refractive index than the central part of the core, so, although they travel a greater distance, it is at a higher velocity. The effect of this is to reduce the differences in the times taken by the many different modes; ideally, all modes then arrive at the distant end in exact synchronism.

No matter which of the three possible types of propagation is used, the dimensions of the outer medium or cladding must be at least several wavelengths. Otherwise some light energy will be able to escape from the system, and extra losses will be caused by any light scattering and/or absorbing objects in the vicinity.

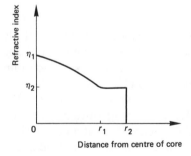

Fig. 23.5
Refractive index profile
of graded-index
multimode optical fibre

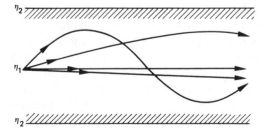

Fig. 23.6
Multimode propagation
in a graded-index fibre

Optic fibre transmission systems have many advantages over "traditional" electrical transmission systems and possess significant characteristics which enable them to provide economical solutions to many telecommunications link requirements:

a) Low transmission loss: this permits longer repeater sections than with coaxial cable systems, thereby reducing costs.

b) Wide bandwidth: this means a large channel carrying capacity.

c) Small cable size and weight: this means that drums of cable can be handled economically (by people instead of by fork lift trucks) and that each cable uses less space in cable ducts.

d) Immunity to electro-magnetic interference: this permits use in noisy electric environments such as alongside electrified railway tracks and means that low signal-to-noise ratios are acceptable.

e) Non-inductive: the fibre does not radiate energy so causes no interference to other circuits. Communications security is thereby enhanced.

f) Long-term cost advantages: the basic raw material, silica, is never likely to be in short supply, and improved technology is continually producing lower cost and more efficient devices.

Optic fibre systems are now in service using bit rates of up to 560 Mbit/s (see section 29); this provides the equivalent of 7680 voice channels carried on a single fibre about the same diameter as a human hair.

24 Fibre to the Curb and to the Home

When Mercury Communications Ltd were first allowed to compete with British Telecom, apart from early provision of service using microwave radio, they relied on a national network based upon a figure-of-eight fibre network between London and the North Midlands, running along British Rail tracks. They provided, and still provide, local network service to their customers in London using fibre running in the abandoned tubes of the old commercial pneumatic tube communications system.

When fibre can provide many hundreds, eventually hundreds of thousands, of voice channels in a single fibre, as we saw in the previous section, why is it economic to provide local network service, where only relatively few circuits are required, using fibre?

The MCL story gives part of the answer. Fibre cables, even cables containing several fibres, are small. They can be drawn into existing ducts even where there is no room left for additional conventional copper cables. Once installed they provide a capacity which will serve future needs for many years to come. It is not economic to over-provide using copper cable to anything like the same extent.

The MCL fibre local networks in London and the other major cities of the UK serve mainly business customers who possess their own private exchanges. This overcomes one big disadvantage of fibre in the local network, that of providing feed current to the telephone on all direct exchange lines. For PABX extension telephones the feed current, required to drive a direct current through the carbon microphone so that our alternating voice signals can be produced as alternating modulation of this direct current, is provided by the PABX. For any single direct exchange line this current has to be provided by the local public exchange and therefore a fibre local distribution network would not be possible.

There is also the cost of fibre, which is presently about ten times the cost of the equivalent copper cable. If, however, the cost is calculated to include the extra new trenches needed to run extra new cable ducts for many more of the copper cables than of the fibre (because the fibre will more often fit in existing ducts), then the cost differential may disappear or even come out in favour of fibre.

We have been using telephones with carbon microphones since the time of Alexander Graham Bell. Is it absolutely essential to have a feed current to the telephone? Well, no it isn't. The ISDN assumes that the digital telephone will often be locally powered. Many existing telephones use local power to operate their many special features. Even if we have to keep the feed current we could provide it over just the last hundred metres of copper from a distribution point in the street or in the basement of the apartment block or office block. For this reason there are two new abbreviations current in the US industry to describe the increasing use of fibre local distribution there—Fibre to the Curb (FTTC) and Fibre to the Home (FTTH).

Indeed, much less is heard of the problem of the local feed in those countries where there is an existing extensive cable television network which is being proposed for local telephone network distribution using ISDN or Broadband ISDN (BISDN). Characteristically, in the UK, British Telecom have been prevented from offering CATV in association with telephone service because of monopoly considerations, whereas the CATV operators are permitted to offer local network telephone service on their existing or planned CATV networks. It is this sort of happening resulting from the liberalization of one-time monopolies that introduces delay in the economic provision of improved services.

On the other side of the coin, British Telecom are offering ISDN service only at much increased tariffs. There is, in many areas, an economic argument to move everybody to ISDN in order to gain the advantages of a fibre local network as well as the manifest advantages of better service and the availability of enhanced features. Of course, any customer using the features would need to be charged the appropriate rate for the service.

Already, where completely new local networks are being envisaged in urban areas, the provision in fibre with ISDN access for all must be considered as an option. One of the authors has been involved in such a study for the devastated business centre of Beirut. There is at present a European team developing interim international standards for fibre-based local networks because of the urgent need to provide new telecommunications infrastructures for East Germany and for other European ex-members of the Warsaw Pact.

25 Primary PCM Systems and Time-division Multiplexing

For a single speech channel an 8-bit PCM word is generated every 125 microseconds (because the sampling rate is 8000 times per second). This complete PCM word can be generated and transmitted so rapidly (in less than 4 microseconds) that any transmission path fed by a single PCM channel would be idle most of the time. By the use of high-speed devices it is possible for PCM words from other channels to be slotted in, to fill this unoccupied time. Each channel is given a designated time-slot, repeated every 125 microseconds.

Time division multiplexing is the procedure by which a number of different channels can be transmitted over a common circuit by allocating the common circuit to each channel in turn for a given period of time, i.e. at any particular instant only one channel is connected to the common circuit. The principle of a TDM system is illustrated by fig. 25.1 which shows the basic arrangement of a two-channel TDM system. In this simple example, analog inputs are considered for clarity. TDM is, of course, normally a means of multiplexing digital signals, such as PCM pulses all having the same amplitude.

The two channels which are to share the common circuit are each connected to it via a channel gate. The channel gates are electronic switches which only permit the signal present on a channel to pass when opened by the application of a controlling pulse. Hence, if the controlling pulse is applied to gate 1 at time t_1 and not to gate 2, gate 1 will open for a

Fig. 25.1
A simple two-channel TDM system; $t_1 = a$ series of pulses occurring at fixed intervals; $t_2 = a$ series of pulses occurring at the same periodicity as t_1 but commencing later by an amount equal to half the time interval

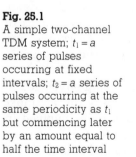
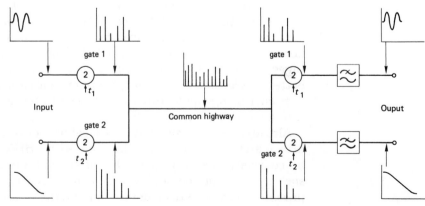

time equal to the duration of the pulse but gate 2 will remain closed. During this time, therefore, a pulse or sample of the amplitude of the signal waveform on channel 1 will be transmitted to line. At the end of the pulse, both gates are closed and no signal is transmitted to line. If now the controlling pulse is applied to gate 2 at a later time t_2, gate 2 will open and a sample of the signal waveform on channel 2 will be transmitted. Thus if the pulses applied to control the opening and shutting of gates 1 and 2 are repeated at regular intervals, a series of samples of the signal waveforms existing on the two channels will be transmitted.

At the receiving end of the system, gates 1 and 2 are opened, by the application of control pulses, at those instants when the incoming waveform samples appropriate to their channel are being received. This requirement demands accurate **synchronization** between the controlling pulses applied to the gates 1, and also between the controlling pulses applied to the gates 2. If the time taken for signals to travel over the common circuit was zero, then the system would require controlling pulses in exact synchronization at both ends, but since, in practice, the transmission time is not zero, the controlling pulses applied at the receiving end of the system must occur slightly later than the corresponding controlling pulses at the sending end. If synchronization signals are sent from one end to the other as an integral component of the PCM system (as they are with internationally specified systems), all the signals will of course maintain their correct relative positions. If the pulse synchronization is correct, the waveform samples are directed to the correct channels at the receiving end. The received samples must then be converted back to the original waveform, i.e. demodulated.

In its passage along a telephone line, the TDM signal is both attenuated and distorted but, provided the receiving equipment is able to determine whether a pulse is present or absent at any particular instant, no errors are introduced. To keep the pulse waveform within the accuracy required, pulse regenerators are fitted at intervals along the length of the line. The function of a pulse regenerator is to check the incoming pulse train at accurately timed intervals for the presence or absence of a pulse. Each time a pulse is detected, a new undistorted pulse is transmitted to line and, each time no pulse is detected, a pulse is not sent.

The simplified block diagram of a pulse regenerator is shown in fig. 25.2. The incoming bit stream is first equalized and then amplified to

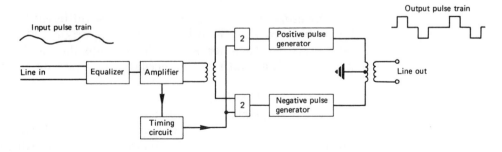

Fig. 25.2
Pulse regenerator

reduce the effects of line attenuation and group-delay/frequency distortion. The amplified signal is applied to a timing circuit which generates the required timing pulses. These timing pulses are applied to one of the inputs of two two-input AND gates, the phase-split amplified signal being applied to the other input terminals of the two gates. Whenever a timing pulse *and* a peak, positive or negative, of the incoming signal waveform occur at the same time, an output pulse is produced by the appropriate pulse generator. It is arranged that an output pulse will not occur unless the peak signal voltage is greater than some pre-determined value in order to prevent false operation by noise peaks.

Provided the bit stream pulses are regenerated before the signal-to-noise ratio on the line has fallen to 21 dB, the effect of line noise on the error rate is extremely small. This means that impulse noise can be ignored and white noise (i.e. noise of constant voltage over the operating bandwidth) is not cumulative along the length of the system. This feature is in marked contrast with an analog system in which the signal-to-noise ratio must always progressively worsen towards the end of the system. Thus, the use of pulse regenerators allows very nearly distortion-free and noise-free transmission, regardless of the route taken by the circuit or its length (fig. 25.3).

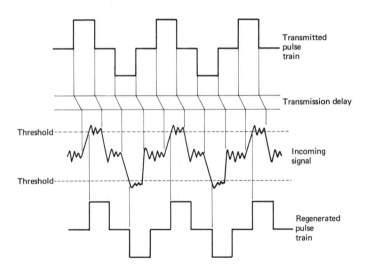

Fig. 25.3
Waveform regeneration
by a pulse regenerator
(*Courtesy*: ITT)

There are, as pointed out in section 13, two different types of PCM, one basically European, the other American in origin. Different quantizing laws are used, together with different signalling and synchronization procedures and a different higher-level multiplexing structure:

"European" (A-law) PCM

Primary multiplexing	30 channels	2.048 Mbit/s
2nd order	120 channels	8.448 Mbit/s
3rd order	480 channels	34.368 Mbit/s
4th order	1920 channels	139.264 Mbit/s
5th order	7680 channels	560.000 Mbit/s

"American" (µ-law) PCM

Primary multiplexing	24 channels	1.544 Mbit/s
2nd order	96 channels	6.312 Mbit/s
3rd order	672 channels	44.736 Mbit/s
4th order	4032 channels	274.176 Mbit/s

Modern PCM system terminals make much use of integrated circuits, many functions being performed in a single chip. To enable principles to be shown, a schematic diagram of an early type of 30-channel PCM is given in fig. 25.4. It will be seen that the 2048 kbit/s line signal is produced in stages:

1) Analog input is fed through an 8 Hz sampling gate, giving in effect PAM or pulse amplitude modulated samples.
2) These PAM samples are multiplexed together on a time-division basis.

(Both these functions are represented by the rotating switch in the diagram.)

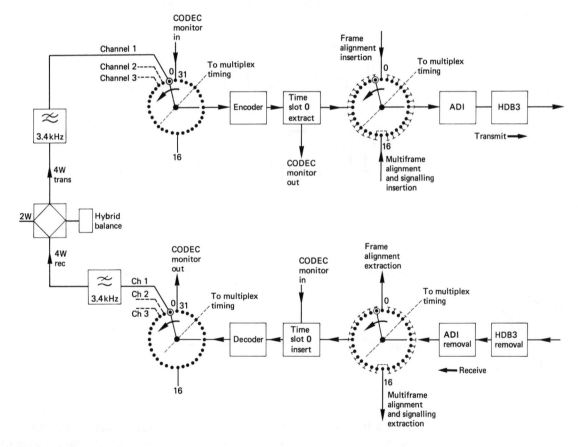

Fig. 25.4
A block schematic of a 30-channel PCM terminal (*Courtesy*: British Telecom)

3) The multiplexed PAM samples are encoded; this quantizes the levels of the samples and prepares the strings of binary digits to indicate these amplitude levels.

4) Synchronization is established.

5) Frames are aligned and signalling inserted.

6) The binary signals representing the samples go through an ADI or Alternate Digit Inversion stage (see fig. 25.5) to overcome some of the disadvantages of unipolar transmission.

7) These ADI signals then go through an AMI or Alternate Mark Inversion stage in order to produce a signal which may be transmitted and regenerated with minimum distortion.

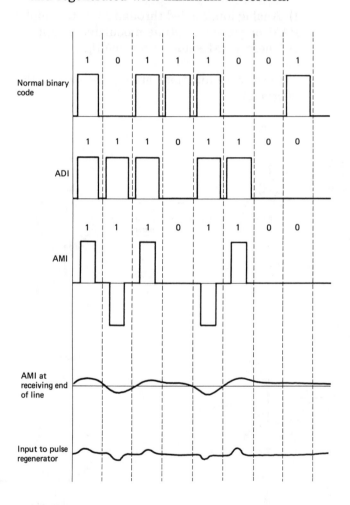

Fig. 25.5
ADI (alternate digit inversion) and AMI (alternate mark inversion) (*Courtey*: British Telecom)

A modified form of AMI called HDB3 (High-Density Bipolar 3) is nowadays used. This overcomes difficulties caused by the fact that ordinary AMI does not code zeros in any way, so if there is a long sequence of consecutive zeros it is difficult to maintain the correct timing relationship between the receive terminal and the received signal. With HDB3, after three successive zeros, the fourth binary zero is replaced by a mark

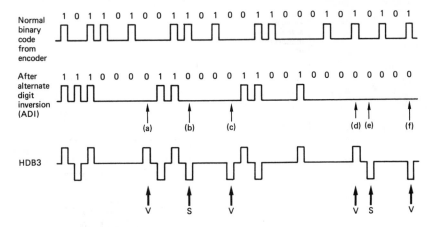

Fig. 25.6
High-density bipolar
code HDB3 (*Courtesy*:
British Telecom)

In the example, 4 zeros are detected by (a), so a violation
mark (V) is inserted. Another 4 zeros are detected by (c),
so a second violation mark is inserted. An additional mark
(S) is inserted at (b). Further violations are detected by (d)
and (f) and an additional mark in inserted at (e) but not
between (c) and (d).

signal of the same polarity as the previous mark signal (see fig. 25.6). Because this is a violation of the alternate mark inversion rules, the receiving equipment knows that this received mark signal is not a genuine mark but represents another zero. If either an even number of marks or no marks at all exist between each violation, then an additional mark (S in fig. 25.6) is substituted in place of the first of the four zeros. The substitution mark has a polarity opposite to that of the previous mark.

26 Integrated Services Digital Networks

An ISDN is a single network able to carry and switch a wide variety of telecommunications services. It can evolve from an IDN, an integrated digital network, which is a telephony network in which digital transmission systems have been fully integrated with digital switching systems (fig. 26.1).

To most people who work in national telecommunications systems and have grown up in countries where telephone services are provided and maintained by a national monopoly, e.g. British Telecom or a PTT Department, it seems fair and reasonable to expect that this national service should be expanded to provide the public with all the new

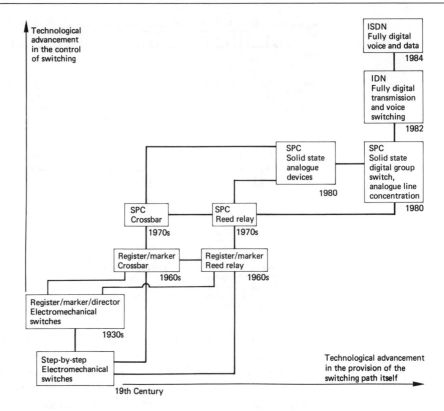

Fig. 26.1
The development of
automatic telephony
leading to the ISDN

telecoms services which are now becoming available, such services as:

Slow-scan television.
Fast high-quality facsimile.
High-quality videotext (Picture Prestel).
Text transmission (teletex).
Computer—computer connections (circuit or packet switching).
Electronic funds transfer.
Telemetry (meter readings for power, water, etc.).
Interactive videotex (electronic shopping).
Electronic mail.
Videophones.
Confravision.

To other people this expectation is not always considered fair and reasonable at all. The computer industry has, for example, developed to its present strength as a largely privately-owned and competitive industry. Equipment and software made by one computer manufacturer will, in general, not work with equipment from other manufacturers. There are very many private data networks in service all over the world; very few of these can be connected to each other or to the public switched network.

With the provision of complex national networks of computer-controlled telephone exchanges capable of switching, not only telephony but also all of these new telecoms services, the "traditional" providers of information processing services, the computer manufacturers, may be forgiven for regarding these new national networks as the beginnings of head-on competition. The convergence of communications and computing has been called information technology (IT) or telematics; as public networks incorporate digital technology features, and as private computing networks enlarge their intercommunication telecoms facilities, it could well be that the resultant information technology revolution will result in our national economies being restructured just as radically as they were in the nineteenth century Industrial Revolution.

The boundaries between communications and computing are being blurred by the move towards digital transmission systems, digital exchanges, digital storage devices and digital common channel signalling capable of controlling the establishment of calls between all the various new telecommunications services.

Subscriber access to the ISDN is planned to be provided at basic level by digital transmission systems carrying bidirectional speech, data and signalling on a single pair of wires in the existing local distribution cable network. The transmission system will allow subscribers the simultaneous use of a number of channels providing:

a) a 64 kbit/s both-way voice or data circuit,
b) a second 64 kbit/s both-way circuit, available to be used either as a second voice circuit or for fast data,
c) a 16 kbit/s data circuit to be used for signalling and relatively slow data.

This is called Basic rate access.

Connection and interworking of a wide range of terminal types will be possible, from a simple telephone instrument to a sophisticated computer or intelligent data terminal.

The arrangement of ISDN basic access is shown in fig. 26.2. Perhaps the most attractive feature of this is that the 144 kbit/s capacity of the line is made available to up to eight terminal devices. These may be telephones, facsimile machines or any form of data device, such as a personal computer. Thus the provision of basic access could remove the need to have a small PABX or key system *and* the need to have a small LAN in many high street shops and offices.

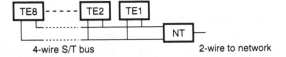

Fig. 26.2
ISDN basic access

At the start of every ISDN call the terminal provides information to the exchange indicating the procedures to be used to set up and control the call. Messages containing information more meaningful to data terminals or computers can replace the tones or verbal announcements which are acceptable to human users. If the service indication is "telephony", the call would not be restricted to the ISDN but could be switched through to anywhere in the worldwide telephone network. If the service indication is for a special service, it may be necessary to provide special interworking interfaces, e.g. to a packet switching network.

It will also be possible to access ISDNs at "primary" level of 2048 kbit/s or 1544 kbit/s, the equivalent of 30 voice circuits on "European" PCM or 24 on "American" PCM systems respectively. This is called Primary rate access.

It should be noted that the USA's AT&T is already thinking further ahead, AT&T say that ISDNs are not a final destination at all, but a route to a bigger future: public networks will eventually evolve into powerful, software-driven networks with highly distributed intelligence called Universal Information Services, UIS.

27 A Word about Protocol

The ISDN, section 26, and the packet switching service, section 35, introduce the concept of a signalling and communication protocol. It would be helpful to examine the concepts involved.

In telecommunications we have always been conscious of the need for signalling codes but only relatively recently, with the introduction of computer communications and of common channel signalling systems, has the term protocol been introduced. The difference between the use of the terms "code" and "protocol" is not always clear.

A reasonable definition of the terms as used in telecommunications is the following:

Code   Any system of symbols and rules for expressing data and instructions in a form understandable by a computer or other equipment for processing or transmitting information.

Protocol   The formula for arranging the parts of a message consisting of coded data and instructions into a form usable and understandable by communicating information processing devices.

We can illustrate the need for a protocol by considering the communication between two people by letter (fig. 27.1). The recipient will not act on the message unless it carries an authorized signature (PASSWORD), confirming that it really comes from the sender. Nor is the recipient likely to act expeditiously unless the letter is couched in acceptably polite terms. In telecommunications these functions are all part of error detection and correction and, perhaps, encryption. Address information is evidently fundamental and the complete message must be packed in a convenient form for the postman to handle. Figure 27.1 also shows how this whole process can be seen as a succession of levels of abstraction

1)	Agree language understood by sender and recipient.	**CODE**	
2)	Compose and write text.	**PROTOCOL**	MESSAGE
3)	Employ polite formulae at beginning and end of message, e.g. Dear Sir ... Yours faithfully Dear Anne ... Yours sincerely		CHECK Redundant information enabling validity of message to be assured.
4)	Signature		PASSWORD SENDER IDENTITY
5)	Address		ADDRESS
6)	Copies to:		BROADCAST ADDRESSES
7)	Envelope and stamp		TRANSMISSION INSTRUCTION

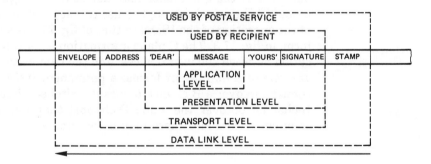

Fig. 27.1
Communication by letter

from the communications link (the postman), each level causing a new outer covering to be wrapped around the central message.

The corresponding diagram showing, in very much simplified form, how a message is transmitted across the ISDN is shown in fig. 27.2. Each descending layer in the hierarchy adds its own header and, perhaps a trailer to the message, and intermediate nodes consult this header information in determining how to pass on the message.

In this description the "envelopes" are information directed at the postman, the communications devices, and have become known as the information for the Control plane, the C-plane The message itself is directed at the communicating users, which may be human users or

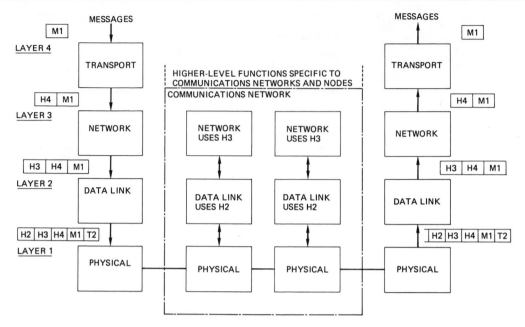

Fig. 27.2
ISDN message
hierarchy

machines; this communication is taking place on the User plane, the U-plane.

The ISDN makes this clear logical separation of signalling information and message information or data. The C-plane determines the routing of the message whereas the message itself is routed via the U-plane. In the basic ISDN the two planes may not have a recognizably separate existence but the principles applied use this separation. Thus, a truer picture of the operation of the ISDN to that of fig. 27.2 is shown, in even simpler form, in fig. 27.3. The C-plane information is used to establish communication between the user devices, and the communication itself may take any one of a number of forms: a permanent (voice or high-speed data) connection over a B-chanel, a packet switched virtual connection over a B-channel or over the same D-channel that was used in the C-plane communication, and there may be others.

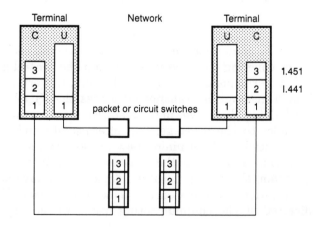

Fig. 27.3
ISDN switched mode
services

28 Broadband ISDN (BISDN)

Hardly had the term Integrated Services Digital Network been invented and described than the industry added a letter and began talking about the Broadband ISDN. It would appear, in retrospect, to be unduly ambitious to conceive a broadband network when the whole of the 1980s and, it seems, most of the 1990s will have been spent in deploying an inadequate ISDN capability.

The suggestion for Broadband first came from the German network provider, at that time the Deutsche Bundespost (DBP). Their approach has been consistently to envisage a broadband ISDN as the ultimate goal and therefore seeing little value in intermediate ISDN service offered to the public over the existing network. To this end, the DBP introduced a broadband trial on the local network, called Bigfon, in 1984. Whereas ISDN has been generally seen as a means of utilizing the existing local network more effectively, the opposing view, typified by the DBP approach, sees the services possible via ISDN, in lacking video, high-speed document transfer, and computer-aided design dialogue, as being quite inadequate. For the Bigfon experiments some 300 subscriber's premises on a new estate in Berlin were equipped with optical fibre distribution providing voice, data, videophone and three simultaneous TV channels to the subscriber. The German strategy was to equip new housing and office developments with optical fibre local distribution as part of the services provision.

Opinions in favour of a rapid transition to a BISDN tend to be most strongly held in countries which have deployed cable TV systems fairly widely. The cable TV network represents an existing wideband local network which constitutes a viable alternative to the local telecommunications network and could possibly provide BISDN access capability. There is a potential conflict here with direct broadcasting by satellite (DBS) (see section 53) which is a direct competitor with cable TV but does not include provision for two-way communications and therefore precludes the provision of interactive services.

Despite this enthusiasm for BISDN in some nations, the publication of the necessary international standards has lagged behind even the prolonged delays for ISDN standards. Additional impetus has now been provided, again by Germany, as a result of German re-unification. There is an urgent need to replace the quite inadequate telecommunications network in former East Germany and this could most easily be achieved by using fibre local distribution throughout. As a result, the European nations have set up a special working group to rush the necessary standards into being to enable the re-wiring of East Germany to be achieved.

Two methods of providing the large transmission capacities of the BISDN in a way that allows flexible allocation of a very wide range of data rates are proposed. The methods are seen as being used in co-operation. One is the Synchronous Digital Hierarchy (SDH), the other is Asynchronous Transfer Mode (see section 29).

The SDH is a means to load a 150 Mbit/s "envelope" with a wide variety of different services at a wide variety of data rates and still be able to multiplex and de-multiplex individual portions without having to break down the complete envelope. This is done by providing a pointer in a known part of the envelope to the whereabouts within the envelope of each individual communication.

The ATM approach is a development of the packet switching method using small, equal-size packets, called "cells", which are small and simple enough to be dealt with at the very high speeds involved.

The overwhelming advantage of the ISDN is the provision of a wide range of services via a single network over a single standard access. Thus the basic access and primary access defined for the ISDN are seen to be fundamental and it is not sensible to add a further set of access interfaces for the BISDN. Clearly new accesses are required, to interface with the 150 Mbit/s SDH at least, and this is the only new access proposed. The BISDN will arrive at the (very large) user and at the local public network distribution point as a 150 Mbit/s SDH transmission. Within the SDH envelope, services requiring the high capacity which the BISDN provides will be in ATM format. "Normal" services, including plain old ISDN will be extracted out of the SDH as 2 Mbit/s systems for presentation at an ISDN primary access or at digital-to-analog conversion equipment for onward analog delivery or 144 kbit/s systems for delivery to an ISDN basic access.

29 Advanced Transmission Methods

The most recent developments influencing digital networks are the provision of high-capacity digital transmission media using fibre optic technology. This has necessitated interest in switching and cross-connecting, not only at the 64 kbit/s and sub-rate levels, but also at the system levels of the higher-order multiplexes. The multiplex hierarchies in use are listed in Table 29.1.

As primary PCM systems, 2 Mbit/s or 1.5 Mbit/s in North America, began to saturate the network, transmission hierarchies were developed to multiplex traffic to higher rates (Table 29.1). Each primary system has its own independent clock which results in slight differences in frequency. This is known as plesiochronous transmission. To compensate for this, padding bits are added at each multiplexing stage. With the introduction of digital switching however, primary rate systems have become largely synchronous to each other but the higher-order bearers are plesiochronous and independent.

	N. AMERICA			REST OF WORLD		JAPAN	
	Code	*Bit rate* (Mb/sec)	*Voice CH*	*Bit rate*	*Voice CH*	*Bit rate*	*Voice CH*
Sub 64		0.0024 0.0048 0.0096 0.056		0.0024 0.0096 0.0192			
Level "0"	DS0	0.064	1	0.064	1	0.064	1
Level 1	DS1	1.544	24	2.048	30	1.544	24
"1+"	DS1C	3.152	48	–		–	
2	DS2	6.312	96	8.448	120	6.312	96
3	DS3	44.736	672	34.368	480	32.064	480
4	DS4E DS4	139.264 274.176	2016 4032	139.264	1920	97.728	1440
5	–	–	–	565.148	7680	397.20	5760

Table 29.1 Digital hierarchies

Although these arrangements were well suited to the transport of bits there remained a major drawback. The confusion of multiplexing different levels with their padding bits made it impossible to identify and extract an individual channel within a high-capacity bit rate link. The complete de-multiplex procedure had to be carried out in order to gain access to a particular channel.

This situation was acceptable when only point-to-point links were involved which were fully terminated at either end. Applications such as drop-and-insert were not viable except, perhaps, at primary rates.

As networks developed and became more complex, the interlinking of traffic at nodes required banks of multiplexers and large distribution frames or digital cross-connects (DCC). Figure 29.1a illustrates the situation. It became clear that the existing multiplexing standards, designed for point-to-point links, were no longer suitable for large meshed networks.

As a result, a number of bodies, including AT&T, Bellcore and British Telecom, developed alternative approaches in which channels were arranged synchronously within the higher-order multiplex. Low-rate channels could then be extracted without going through the de-multiplexing process. Figure 29.1b illustrates what this might do to the network.

Synchronous Digital Hierarchy (SDH)
New proposals, based on this perception, were driven initially by North American requirements and were expressed in the North American

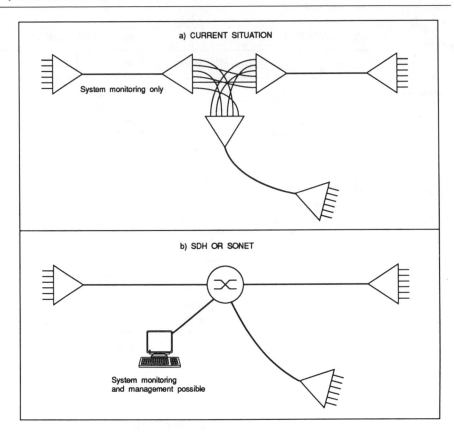

Fig. 29.1
Development into
synchronous networks

standards for Synchronous Optical Network (SONET). In February 1988, however, agreement was reached at CCITT for a world standard. This standard is published in the Blue Book as Recommendations G.707, G.708 and G.709 under the heading Synchronous Digital Hierarchy (SDH), a title which replaces SONET.

The SDH standard has, or ought to have, four important features:

1) It is synchronous which allows efficient drop-and-insert and cross-connect applications.

2) It is an optical standard ensuring compatibility at the optic signal level between equipment from different manufacturers.

3) It makes provision for network management channels ensuring effective control and reconfiguration of networks.

4) It can be introduced directly into existing networks in both the 24-channel and 30-channel "worlds".

It is being formulated at the same time as the BISDN requirements and can embrace these within the SDH standards.

The philosophy underlying the standard is that any of the currently used transmission rates can be packaged into a standard-sized container and located in an easily identifiable position in the multiplex structure.

For all the transmission rates in use there exist, or are at least proposed, mappings into containers, called Virtual Containers (VC). Once located in containers, multiples of different containers can be combined together into a standard format. In this way the structure can be used for 2 Mbit/s, 8 Mbit/s, 34 Mbit/s and 140 Mbit/s or for North American 1.5 Mbit/s, 6 Mbit/s and 45 Mbit/s traffic.

Containers can be mixed allowing 1.5 Mbit/s and 2 Mbit/s traffic to be carried simultaneously within the same structure.

To achieve efficient mapping within a 140 Mbit/s structure and to provide sufficient management overhead, 155 Mbit/s has been chosen as the basic SDH rate. This is known as Synchronous Transport Module 1 (STM1) and forms the building block for higher-rate traffic. Higher rates are formed by simple multiplexing (Table 29.2).

STM 1	155 Mbit/s
STM 2	300 Mbit/s
STM 4	600 Mbit/s
STM 8	1.2 Gbit/s
STM 16	2.4 Gbit/s

Table 29.2 SDH Rate Structure

A further level of control is provided for network administration. The Administrative Unit (AU) is the basic unit through which network operators can control the network. Virtual containers are selected for amalgamation into an AU, or "groomed", prior to the AUs themselves being combined into an STM. This grooming procedure can sort individual channels according to their outgoing destinations.

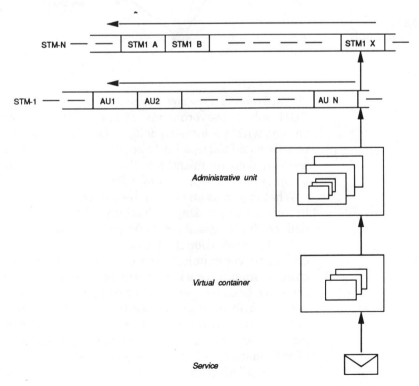

Fig. 29.2
Principle of
synchronous digital
hierarchy

Figure 29.2 illustrates the SDH levels but does not show the particular methods, such as byte interleaving, used to multiplex to the higher orders.

It ought now to be fairly clear to the reader that implementation of SDH will allow high-capacity networks to develop without the need for complex reiterative de-multiplexing processes but using instead combinations of DCC and synchronous intelligent multiplexers to achieve the same effect. Such a network arrangement is shown in fig. 29.3.

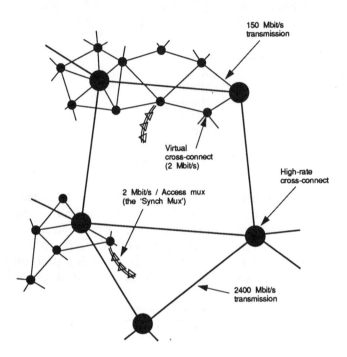

Fig. 29.3
Possibilities of SDH
networks

Asynchronous Transfer Mode (ATM)
SDH solves the problems of transporting many different communications within a high-capacity network by packing the communications together and using a fairly sophisticated pointer mechanism to identify individual communications. The pointer mechanisms however only have to operate at low speeds of a few kbit/s and not at the interface rate.

What began as an alternative method of performing the same task has become a co-operating technology which, in conjunction with the SDH, delivers high-capacity communications over the BISDN.

At the conventional data rates of the ISDN and existing facilities we could only communicate in real time, say by voice, by using circuit switching to guarantee a permanent path. Packet switching was not suitable for voice because the delays involved would destroy the conversation. ATM is a form of packet switching using small equally-sized packets, called cells, and a simple enough protocol to permit cells to be transmitted, interpreted and delivered fast enough for them to carry any kind of information including voice and video.

Using ATM, a wide range of channel rates can be accommodated on a single bearer. The control mechanism must, however, operate at the bearer rate. ATM uses short fixed-length cells with minimal headers to allow calls to be routed at high speed by means of hardware-implemented routing tables at each switching node. International agreement is that the header of each cell will consist of 5 eight-bit bytes and that the cell information field will consist of 48 bytes making a total length of 53 bytes. The header consists of bits to provide two main routing functions.

a) Virtual Path Identifier (VPI). A path is the equivalent of a route in a circuit switched system, permanently connecting two points together. In an ATM environment, the path would not have a fixed capacity. By "virtual" is meant that cells can be routed from source to destination on the basis of the VPI in whatever way seems appropriate.

b) Virtual Call Identifier (VCI). Calls set up as required over the virtual path indicated by the VPI.

ATM can flexibly handle all types of traffic. Examples are:

Voice 64 kbit/s voice can be assembled into ATM cells. With 48 bytes of information in each cell, the cell can contain 6 ms of speech. For some purposes the delay this introduces may be excessive in which case only partially filled cells may be used.

Video Using modern techniques of condensed video, only changes of the picture signal are transmitted. This variation with the activity in the picture is ideally suited to the variable capacity features of ATM.

Signalling Signalling is based on the existing ISDN protocols assembled into cells. Some of these cells would have to be given high priority necessitating the use of flow control techniques in ATM.

Part E
Radio Systems

30 Radio Propagation

1 The Ionosphere

Ultra-violet radiation from the sun entering the atmosphere of the earth supplies energy to the gas molecules of the atmosphere. This energy is sufficient to produce ionization of the molecules, that is remove some electrons from their parent atoms. Each atom losing an electron in this way has a resultant positive charge and is said to be ionized.

The ionization thus produced is measured in terms of the number of free electrons per cubic metre and is dependent upon the intensity of the ultra-violet radiation. As the radiation travels towards the earth, energy is continually extracted from it and so its intensity is progressively reduced. The liberated electrons are free to wander at random in the atmosphere and in so doing may well come close enough to a positive ion to be attracted to it. When this happens, the free electron and the ion recombine to form a neutral atom. Thus a continuous process of ionization and recombination takes place.

At high altitudes, the atmosphere is rare and little ionization takes place. Nearer the earth the number of gas molecules per cubic metre is much greater and large numbers of atoms are ionized; but the air is still sufficiently rare to keep the probability of recombination at a low figure. Nearer still to the earth, the number of free electrons produced per cubic metre falls, because the intensity of the ultra-violet radiation has been greatly reduced during its passage through the upper atmosphere. Also, since the atmosphere is relatively dense, the probability of recombination is fairly high. The density of free electrons is therefore small immediately above the surface of the earth, rises at high altitudes, and then falls again at still greater heights. The earth is thus surrounded by a wide belt of ionized gases, known as the ionosphere.

In the ionosphere, layers exist within which the free electron density is greater than at heights immediately above or below the layer. Four layers exist in the daytime (the D, E, F_1 and F_2 layers) at the heights shown in fig. 30.1.

The heights of the ionospheric layers are not constant but vary both daily and seasonally as the intensity of the sun's radiation fluctuates. The electron density in the D layer is small when compared with the other layers. At night-time when the ultra-violet radiation ceases, no

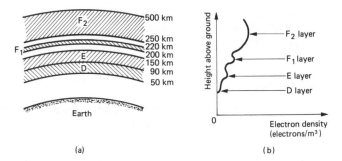

Fig. 30.1
Layers in the
ionosphere

more free electrons are produced and the D layer disappears because of
the high rate of recombination at the lower altitudes. The E layer is at
a height of about 100 km and so the rate of recombination is smaller.
Because of this, the E layer, although becoming weaker, does not normally disappear at night-time. In the daytime, the F_1 layer is at a more
or less constant height of 200–220 km above ground but the height of
the F_2 layer varies considerably. Typical figures for the height of the F_2
layer are 250–350 km in the winter and 300–500 km in the summer.

The behaviour of the ionosphere when a radio wave is propagated
through it depends very much upon the frequency of the wave. At low
frequencies the ionosphere acts as though it were a medium of high electrical conductivity and reflects, with little loss, any signals incident on
its lower edge. It is possible for a VLF or LF signal to propagate for considerable distances by means of reflections from both the lower edge of
the ionosphere and the earth. This is shown by fig. 30.2. The wave
suffers little attenuation on each reflection and so the received field
strength is inversely proportional to the distance travelled.

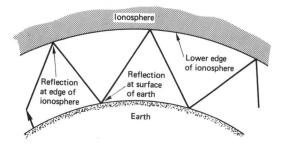

Fig. 30.2
Multi-hop transmission
of a low-frequency
wave

For radio signals in the MF band the D layer acts as a very lossy
medium; MF signals suffer so much loss in the D layer that little energy
reaches the E or F layers. At night-time, however, the D layer has disap-
peared and an MF signal will be refracted by the E layer and perhaps
also by the F layer(s) and returned to earth.

With further increase in frequency to the HF band, the ionospheric
attenuation falls and the E and F layers provide refraction of the sky
wave. At these frequencies the D layer has little, if any, refractive effect
but it does introduce some losses.

The amount of refraction of a radio wave that an ionospheric layer is able to provide is a function of the frequency of the wave, and at VHF and above useful refraction is not usually obtained. This means that a VHF or SHF signal will normally pass straight through the ionosphere.

2 Fading

Fading, or changes in the amplitude of a received signal, is of two main types: general fading, in which the whole signal fades to the same extent; and selective fading, in which some of the frequency components of a signal fade while at the same time others increase in amplitude.

As it travels through the ionosphere, a radio wave is attenuated, but since the ionosphere is in a continual state of flux the attenuation is not constant, and the amplitude of the received signal varies. Under certain conditions a complete fade-out of signals may occur. General fading can usually be combated by automatic gain control (a.g.c.) in the radio receiver.

The radio waves arriving at the receiving end of a sky-wave radio link may have travelled over two or more different paths through the ionosphere (fig. 30.3a). The total field strength at the receiving aerial is the phasor sum of the field strengths produced by each wave. Since the ionosphere is subject to continual fluctuations in its ionization density, the difference between the lengths of paths 1 and 2 will fluctuate and this will alter the total field strength at the receiver. Suppose, for example, that path 2 is initially one wavelength longer than path 1; the field strengths produced by the two waves are then in phase and the total field strength is equal to the algebraic sum of the individual field strengths. If now a fluctuation occurs in the ionosphere causing the difference between the lengths of paths 1 and 2 to be reduced to a half-wavelength, the individual field strengths become in antiphase and the total field strength is given by their algebraic difference.

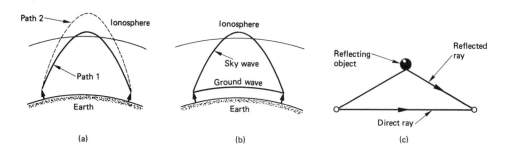

(a) (b) (c)

Fig. 30.3
Multi-path propagation

The phase difference between the field strengths set up by the two waves is a function of frequency and hence the phasor sum of the two field strengths is different for each component frequency in the signal. This means that some frequencies may fade at the same instant as others are augmented; this can result in distortion of received signals.

Selective fading cannot be overcome by the use of a.g.c. in the receiver but several methods of reducing it do exist. For example, the use of frequencies as near to the maximum usable frequency as possible, the use of a transmitting aerial that radiates only one possible mode of propagation, the use of single-sideband or frequency-modulated systems, or the use of such specialized equipment as Lincompex (linked compression and expansion). Selective fading of the sky wave is most likely when the route length necessitates the use of two or more hops. Suppose, for example, that a two-hop link has been engineered. Then, because of the directional characteristics of the transmitting aerial, there may well also be a three-hop path over which the transmitted energy is able to reach the receiving aerial.

Selective fading can also arise with systems using both surface and space waves. In the daytime the D layer of the ionosphere completely absorbs any energy radiated skywards by a medium-wave broadcast aerial. At night the D layer disappears and skywards radiation is returned to earth and will interfere with the ground wave, as shown in fig. 30.3*b*. In the regions where the ground and sky waves are present at night, rapid fading, caused by fluctuations in the length of the sky path, occurs.

Figure 30.3*c* illustrates how multi-path reception of a VHF signal can occur. Energy arrives at the receiver by a direct path and by reflection from a large object such as a hill or gas holder. If the reflecting object is not stationary, the phase difference between the two signals will change rapidly and rapid fading will occur.

Similar phase differences and rapid fading will occur in mobile systems while the mobile unit is moving in towns and receiving signals which are reflected from buildings. This is a problem which has to be considered in the design of mobile systems and receivers.

31 Antennas

In a radio communication system the baseband signal is positioned in a particular part of the frequency spectrum using some form of modulation. The modulated wave is then radiated into the atmosphere in the form of an electro-magnetic wave by a transmitting antenna (or aerial). A transmitting antenna may handle many kilowatts of power and has to be carefully matched to its feeder cable to ensure maximum power input.

For a radio signal to be received at a distant point, the electromagnetic wave must be intercepted by a receiving antenna. A receiving antenna may only be concerned with a few milliwatts of power; its priority is usually for maximum gain and directivity.

The common types of radio antenna include:

1) Dipole, of which the Yagi and the Curtain are developments
2) Rhombic
3) Log-periodic
4) Parabolic

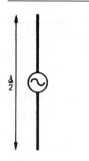

1) *The Dipole* is basically a conductor whose electrical length is one-half the wavelength at the desired frequency of operation, and is centre-fed. This is the basic lambda/2 dipole, shown in fig. 31.1.

The radiation patterns, or the graphical representation of the way in which the electric field strength produced by an antenna varies at a fixed distance from the antenna, in all directions in the plane concerned, are given in fig. 31.2 for the horizontal plane (a circle) and for the vertical plane.

Fig. 31.1
The λ/2 dipole

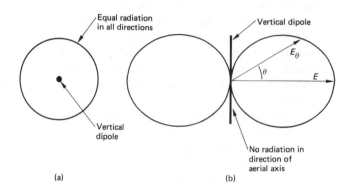

Fig. 31.2
Radiation patterns of a vertical λ/2 dipole:
(a) horizontal plane pattern;
(b) vertical plane pattern

One development of the dipole is to use multiple dipole elements in either curtain or co-linear arrays. A further development, the Yagi, named after its Japanese inventor, uses the principle of increasing the directivity of the lambda/2 dipole by adding parasitic elements called reflectors (about 5% longer than the dipole) and directors (about 5% shorter), mounted at carefully calculated distances from the dipole itself (figs 31.3 and 31.4). More directors, all mounted on the side of the dipole facing the direction of transmission, will give more directivity. Yagi antennas are widely used for the reception of TV broadcast signals (fig. 31.5).

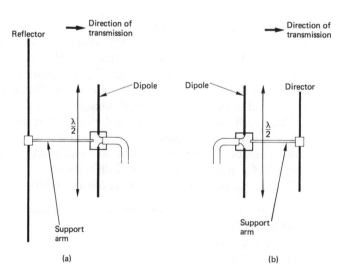

Fig. 31.3
λ/2 dipole with
(a) a reflector and
(b) a director

Fig. 31.4
Radiation patterns: (a)
λ/2 dipole and reflector,
in equatorial plane; (b)
λ/2 dipole and reflector,
in meridian plane; (c)
λ/2 dipole, reflector and
director, in equatorial
plane

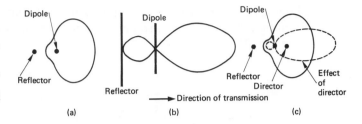

Fig. 31.5
A practical Yagi aerial

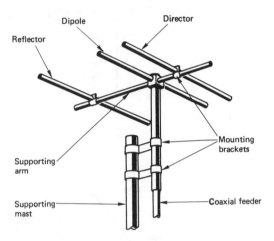

2) *The Rhombic*   A rhombic antenna, because it is used at HF frequencies, lower than the VHF and UHF where Yagis are utilized, is very much larger than a Yagi. It can take up more area than a football field and is used principally for point-to-point HF links. The lengths of each arm and the height above the ground all have to be calculated for optimum efficiency at the particular frequencies to be used and the distance and bearing of the receiving station. The angle of elevation of the main beam (determined largely by the height of the antenna) is important because rhombics are used primarily with sky-wave propagation systems.

3) *The Log-periodic*   These are much used for point-to-point services. They do not give quite so much directional gain as rhombics but they take up much less land area. They are able to operate efficiently over wide frequency bands; variants can be used for VHF or HF services. A log-periodic antenna is made up of a series of rod or wire dipoles with a common ratio both between the lengths of adjoining rods or wires and for the spacings between them. Figure 31.6 shows typical VHF and HF log-periodic antennas.

4) *The Parabolic Reflector*   Frequencies at the upper end of the UHF band and in the SHF band can be treated in much the same way as light beams. Just as a parabolic reflector is used in a searchlight to produce a powerful parallel beam of light when the light source is located exactly at the parabolic focus, so a parabolic reflector (often called a dish) can be used to provide a very directional high-gain antenna with the radio energy concentrated into a parallel beam. Dish diameters vary from

about 20 cm for an internal antenna for a domestic TV receiver up to more than 30 metres for ground stations working to geostationary satellites (fig. 31.7).

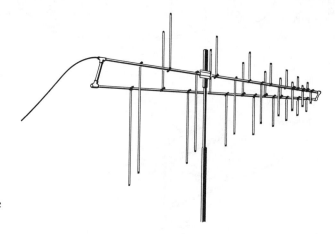

Fig. 31.6
A practical log-periodic aerial

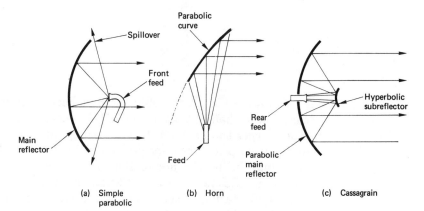

Fig. 31.7
Three versions of the parabolic antenna, showing placement of RF feed and the path of reflected energy. The Cassegrain is the most common for satellite communications

32 Satellites

The first man-made satellites to orbit the earth travelled at altitudes of only a few hundred miles; they went round the earth in about an hour. Nowadays most communications satellites are placed in a geostationary orbit at 22 240 miles (35 780 km) above the earth in the equatorial plane. At this altitude the orbit takes 24 hours so that the satellite appears from the earth to be stationary. However satellites in other orbits ore also in use and there is a rising interest in low earth orbits (LEO).

Travelling at the speed of light the radio signals between the satellite and the ground suffer a delay of between 0.2 and 0.4 of a second. It is therefore necessary to limit the number of "hops" in a satellite connection where the delay would adversely affect the communication. This is particularly true for voice communications.

There are three international organizations with civil communications satellites in orbit: Intelsat, Intersputnik and Inmarsat. The role of Inmarsat is described in section 33, Maritime Communications. Other satellite systems are owned and run on a regional or country basis. Notable of the regional systems are Eutelsat for Europe and Arabsat for the Middle East. Among the countries operating their own satellites are Indonesia (Palapa), Australia (Ausat), Brazil (Brazilsat) and India (Insat). All these satellites carry a number of transponders which may be used for multiple telephone or data circuits or for television.

A satellite link consists of three equipments: an earth station at the sending end, the satellite, and the earth station at the receiving end. The two earth stations will often be in different countries and will each be owned by the telecommunications operator in the country, while the satellite will be owned and operated by another organization with a Tracking, Telemetry and Control (TTC) station probably in yet another country.

There are currently (1992) 122 members of the Intelsat organization, Russia and Azerbaijan being the latest two to join. There are more than 1300 earth stations accessing the Intelsat system providing more than 2200 communication links worldwide. The satellites are also used for domestic communications by some of the member countries.

Intelsat currently has 15 satellites in orbit and these are arranged to cover the earth in three regions: Atlantic Ocean Region, Indian Ocean Region and Pacific Ocean Region. Although it may appear strange to divide the system according to the oceans, it has to be remembered that satellites are most useful in spanning large distances.

Of the 15 satellites, 12 are of the Intelsat V or VA series and 3 are of the Intelsat VI series. Seven of them are located over the Atlantic Ocean Region and four each over the Indian and Pacific Ocean Regions. There is a continuous programme of development and upgrading of these satellites as may be understood from the use of the sixth (VI) design at present. Orders have already been placed for satellites of the seventh (VII) series and these are due to start being launched in 1993. Each series is more advanced than the last and the VII series will have more on-board facilities than previous satellites.

The Intelsat V satellites have a capacity for up to 15 000 circuits plus 2 television channels. The Intelsat VI have a capacity of up to 24 000 circuits and 3 television channels while the Intelsat VII series will have a capacity of 18 000 circuits and 3 television channels. The capacity may be increased by the use of circuit multiplication techniques, such as the Digital Circuit Multiplication Equipment (DCME) which provides a five times increase in circuit capacity.

The Intelsat VI series of satellites are large spin-stabilized cylinders 3.6 m in diameter and 5.3 m high (before deployment), with solar cells all around the outside of the cylinder to provide power. The antennas are mounted at one end on a "de-spun" shelf to ensure that they always point towards the earth. The Intelsat V and VII series of satellites are of the three-axis-stabilized type with solar panels and antennas which fold out from the satellite after it is in orbit. The V series is 2 m in diameter and 6.4 m high, while the VII series is 2.7 m in diameter and 4.2 m high, each before the panels are deployed.

Intelsat have laid down a number of standards for earth station antennas ranging from Standard A for antennas of 15–18 m diameter used at major earth stations to Standard E1 for antennas of 3.5–4.5 m diameter for the Intelsat Business Service (IBS). There are 14 different standard antennas for various applications. Their characteristics are defined by a minimum gain/noise temperature figure (G/T in dB/°K).

Communications through the satellites may be either analog or digital. They range from a single analog voice circuit using a Single Channel Per Carrier (SCPC) system to multiple digital voice circuits using the Intermediate Data Rate (IDR) system, which may carry up to an 8 Mbit/s digital data stream. Not all communications are between two end users; broadcast mode may be used to send information to many recipients.

The frequencies used for satellite communications are either C-band or Ku-Band. In C-band the uplink uses frequencies in the 6 GHz band and the downlink frequencies in the 4 GHz band. In Ku-band the corresponding frequencies are 14 GHz and 11 GHz.

The Intersputnik system is similar to the Intelsat system. The member states are from Eastern Europe and associates of the former Soviet Union such as Cuba and (South) Yemen. Intersputnik leases capacity on two Gorizont satellites assigned to the Statsionar 4 and 13 orbital positions. 12 earth stations access the Atlantic Ocean Statsionar 4, and 7 the Indian Ocean Statsionar 13. As with Intelsat the traffic consists of video, voice and data.

One of the developing areas of satellite communications is the use of VSATs (Very Small Aperture Terminals) for data connections usually between computers or between a computer and a terminal. VSATs use antennas about 1.8 m diameter, and may carry data up to about 19.2 kbit/s. However the antenna size depends on the data rate which can vary according to the user. This system has become very popular in the USA due to a favourable licensing regime adopted by the Federal Communications Commission (FCC). Such businesses as the "7-Eleven" convenience stores use the system for the verification of credit card ratings and inventory control. The use of VSATs has not become so popular yet in Europe due to a more restrictive regulatory environment and probably due to the shorter distances involved providing a more competitive terrestrial system.

Geostationary satellites are situated above the equator and cannot therefore provide satisfactory communication to the polar regions. The Arctic Circle is an important area to cover for Russia and therefore another system of satellites has been developed to cover this region. These are the "Molniya" satellites. Instead of maintaining a geostationary position, these satellites travel along an elliptical orbit. This orbit takes each satellite from one apogee (high point) 20 000 miles over Russia to a perigee near Antarctica and then back up to a second apogee over Canada. During the period when the satellites are at the apogees they appear relatively stationary and can be used for communications.

Each satellite can be used in this way for about 6 hours at a time. By having several satellites in this type of orbit continuous communications can be maintained.

There is also interest in providing worldwide coverage for mobile communications. The use of low earth orbit (LEO) satellites is proposed for this, since the distance from the mobile earth station to the satellite can be reduced to a few hundred miles by this means. Satellites in such orbits, however, pass overhead quite quickly, so that a number of satellites are needed to provide coverage at any place. One of the systems proposed (by Motorola) is known as Iridium since it uses 77 satellites, the same number as the number of orbiting electrons in an atom of the element Iridium. This system is planned to start operation in 1995. The World Administrative Radio Conference (WARC) held in February 1992 allocated frequencies for the use of this service.

33 Maritime Communications

Communications with ships at sea has been at the forefront of radio communications since the days when Marconi conducted some of his first experiments between ship and shore. While communications are important for us all, for ships they are particularly important for safety and distress purposes. Radio was first used to save life at sea in 1899 when a lightship reported that the steamer Elbe had run aground.

In 1912, soon after the Titanic disaster, an international conference standardized distress and calling frequencies, and made provision for these frequencies to be monitored both at sea and on shore. These frequencies (500 kHz, 8 MHz, 12 MHz, 16 MHz and 22 MHz) are still monitored constantly, often by automatic equipment. In 1914 the International Convention for the Safety of Life at Sea (SOLAS) required that ships within certain categories should carry radio-telegraph equipment, a requirement that has been updated regularly since.

For many years communication was confined to Morse transmissions on short and medium waves. Medium wave was used when the ship was within range of a coast station and short wave when medium wave transmission was not possible. For communications to ships, the location of the nearest coast stations had to be determined before the message could be sent. More recently radio-telephony has been introduced, and telex.

Each coast station transmits a "traffic list" of stations for which messages are held awaiting transmission. The traffic list is transmitted regularly, say every hour. Ship's operators are expected to listen out and, if they spot their own call sign in one of these lists, they are expected to call the coast station concerned and arrange for the message to be transmitted.

Communications are not confined to safety messages, but include normal messages, particularly instructions from ships' owners to their captains, and reports of the positions of ships with their expected arrival date in port.

Short wave transmission, as described in section 3, provides worldwide capability, but requires the use of different frequencies at different ranges, at different times of the day, at different seasons and at different parts of the sunspot cycle. It is also subject to periodic electromagnetic disturbances. The task of the radio operator is thus quite skilled.

With the introduction of radio-telephone and radio-telex, it has been possible to link ships through the coast stations into the public telecommunications network.

When ships get within about 40 km of a major port, they can usually establish contact using a VHF radio-telephone to make arrangements for their arrival and unloading. When they actually dock, their radio apparatus is usually sealed to prevent it being used in port. The ship is then expected to use the local telecommunication services for its communications. This is probably mainly to prevent ships avoiding payment to the local telecommunications authorities, but may also allow censorship to be applied in some circumstances.

Maritime communications has been transformed through the use of satellites, and particularly through the setting up of the International Maritime Satellite Organization (INMARSAT) in 1979. Operational services were started in 1982, and Inmarsat now provides first-class telecommunication services 24 hours a day, 7 days a week to all suitably equipped ships in all the oceans of the world.

While Intelsat divides the world into three (section 32), it is divided into four regions by Inmarsat: Atlantic Ocean Region East, Atlantic Ocean Region West, Indian Ocean Region, and Pacific Ocean Region (see fig. 33.1). The organization has headquarters in London, where the Operations Control Centre is also located. Each region has a Network Coordination Station.

A ship's earth station is a comparatively small affair quite unlike its terrestrial cousins with their 30 meter diameter dishes. Frequencies used are 1.5 GHz downlink from the satellite to the ship, 6 GHz uplink. The coast earth station uses a 4 GHz downlink and a 6 GHz uplink. The ship's antenna is mounted on a sophisticated stabilization system in order to keep the dish pointing straight to the particular satellite concerned however violently the ship rolls, and the whole assembly is covered by a lightweight waterproof dome. Typical dishes are only 0.9 m in diameter. The whole dome unit is less than 2 m high and 1.5 m in diameter, and in weighing less than 200 kg can easily be mounted on the ship's mast out of everyone's way, even on quite a small ship. Many thousand ocean-going ships, yachts and motor launches are equipped with stations of this type.

Initially Inmarsat used rented satellite facilities from Intelsat, the European Space Agency and others, but now builds and launches its own satellites, with four satellites of the Inmarsat II series now in orbit.

Setting up a call to a passenger on a big ship, or to the captain or crewman on a small ship, is, nowadays, almost the same as dialling one's own local neighbour. The CCITT (International Consultative Committee for Telephony and Telegraphy) has drawn up an internationally agreed numbering plan with digits to indicate the country of registration of the

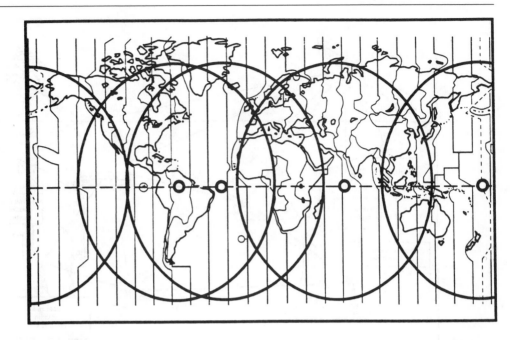

Fig. 33.1
Overlapping
INMARSAT global
beam coverage areas
Courtesy: The World
Satellite Almanac

ship as well as the ship itself, because someone, somewhere on land, has to be willing to be responsible for the allocation of the call charges. Procedure is just as for a normal international self-dialled call.

Now that reliable good-quality voice-grade circuits can be established by satellite to a ship, all the many telecommunications services and terminals which are already in use in offices can be made available in ships, including telex, teletex and facsimile. The availability of up-to-date weather maps on board ship is of great importance to mariners. Developments in this field are continuing, with a service from the World Meteorological Office (WMO), already introduced for airports, being possibly made available through the satellite system.

The total field of maritime communications, including satellite, short wave (HF), MF and VHF, has been integrated into the Global Maritime Distress and Safety System (GMDSS), which ensures that the most appropriate mode of communication is available to a ship at any time (see fig. 33.2). The system includes the provision of Rescue Control Centres (RCC) to deal with any emergency which may arise, possibly involving aircraft either as part of the emergency or assisting. Ships are also provided with portable free-floating emergency position-indicating radio deacons (EPIRB) which are self-activating in the event of an emergency and whose signal is received through polar-orbiting satellites of the COSPAS-SARSAT system.

Inmarsat has now extended the system for use with land-based mobile stations using the "Standard C" terminal, about the same size as an ordinary car radio and a flush-mounted omni-directional antenna, resulting in reliable 600 bit/s text transmission and reception. This is particularly attractive to long-distance lorry operators. Telephone calls can be made using the "Standard M" terminal which provides voice-grade digital calls using a briefcase-sized terminal and a collapsible antenna, a boon to roving reporters.

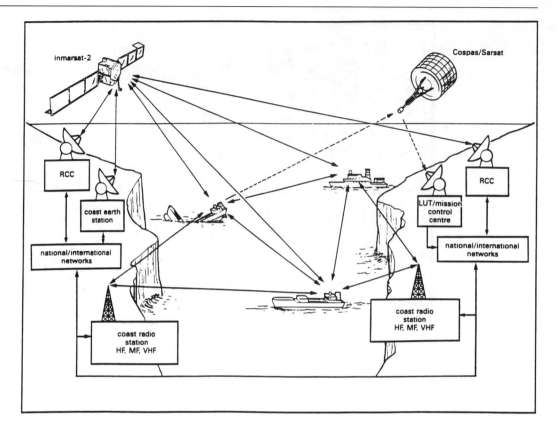

Fig. 33.2
The general concept of
Global Maritime
Distress and Safety
System (GMDSS)
Courtesy: ITU
Telecommunications
Journal, Vol. 59, 1, 1992

A further interesting development of the Inmarsat system is its use for aircraft. Like ships, aircraft travelling over large oceans have had to use short wave radio to communicate, and the air traffic control system has depended on these communications. Several airlines have now fitted some of their fleet with satellite facilities which provide a much more reliable system and can be used not only by the crew but also by passengers. The first of these was the Skyphone system brought into service in 1991 using terminals in the UK (Goonhilly), Norway and Singapore. Other services are now also available.

It is interesting that the Inmarsat earth stations are known as "Coast Earth Stations" in the maritime service and "Ground Earth Stations" in the aeronautical service.

Another recent development of use to maritime and other navigators is the introduction of the Global Positioning System (GPS). This is based on a series of orbiting satellites and is capable of providing positional accuracy of about 20 m. However, the system is managed by the US Air Force, and due to a system of degrading, the absolute accuracy available for civilian use is limited to about 100 m. A system of differential GPS (DGPS) can be used once one known position is established to provide greater local accuracy. The full system of 24 satellites (21 operating and 3 spare) is expected to be in service by the end of 1993. The system had an enormous boost during the Gulf War when it proved its accuracy in the desert.

Part F
Data Systems

34 OSI: Open Systems Interconnection

While the telecommunications network providers were puzzling over the problems of communicating communications processors and solving these problems with common channel signalling, the computer manufacturers and users puzzled over the exactly analogous problem of communicating computers. In the latter case there was more rivalry and more urgency. The problem appeared at times to simplify into the (computer) world seeking to communicate with IBM. The non-IBM world invoked the International Standards Organization and the OSI model was the result. IBM, on the other hand, introduced a similar model, Systems Network Architecture (SNA), and the resolution of OSI and the SNA into a single, universal model is only now nearing achievement.

Where the idea of layering, or enveloping, originated is not clear but it is fundamental to both CCS and OSI although the correspondence of the layers in CCS and OSI is not one-to-one.

When the telecommunications industry addressed the problem of the ISDN, the OSI model was well established and it was both natural and right to organize the ISDN in accordance with it. The ISO definition of the Open Systems Interconnection (OSI) model is contained in ISO 7498. The CCITT has republished the definition without change but alongside considerable additional material applicable to telecommunications. The basic CCITT reference is Recommendation X.200.

An important concept of the OSI model is that of the open system itself. An open system is a processor (with its associated software and peripherals) that obeys standards based on the OSI model. By obeying these standards, open systems can communicate with each other in a network regardless of who made them or programmed their software or set up the network. Interoperability does not require open systems in the strict OSI definition of the word, but open systems of some kind are certainly a great help.

As we have said, the OSI model uses the concept of layers (see section 27). Network activities are seen as taking place in different layers, from transmission of the physical signal at the bottom to the issuing of a network command or the display of a human language message at the top.

The model defines the functions that are to take place in each layer. The lower layers are considered subordinate to the upper layers; each layer makes use of the services of the layer below in performing its function and hands the completed task on to the layer above for further processing. Within a system a layer communicates only to the layer directly above and the layer directly below. Between systems, a layer communicates with the corresponding layer in the other system (peer-to-peer communication). Figure 34.1 and fig. 34.2 attempt to make all this clear.

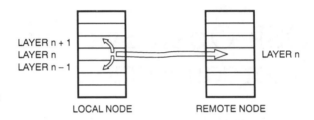

Fig. 34.1
Layered communication
in the OSI model

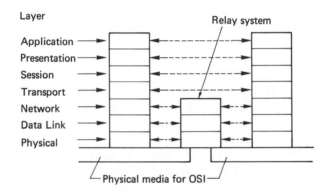

Fig. 34.2
Open systems
communications
showing relaying
through, for example,
switching nodes

In essence, the concept of layers results in a clear division of labour in network processing. The software or hardware that performs the processing at a given layer doesn't care who or what is doing the processing at the next layer above or below. It just does what it is supposed to do when data arrives from one layer and then sends the processed data along to the next layer.

The function of each of the various layers is defined, in rough outline, as follows:

Layer 1 The *physical layer*: specifies the actual physical medium used for the network, coaxial cable, fibre, copper pair. It also defines other aspects of the physical transmission of the data, signal levels and the distance the signal can travel before being regenerated.

Layer 2 The *data link layer*: controls access to the physical medium. This might require sequencing data that has to be segmented upon transmission and re-assembling the data upon reception. This layer also does low-level error detection and correction.

Layer 3 The *network layer*: responsible for routing across the network.

The route from one user to another might require going via other computers or communications switches. Processing in this layer ensures that higher layers do not have to become involved at these intermediate nodes.

Layer 4 The *transport layer*: in charge of maintaining reliable communications across the network. It performs error recovery and regulates the flow of data so that a mainframe does not send more data than a PC can handle for example. The transport layer is the first of the end-to-end layers; it is not concerned how the data arrives at its destination, only that it is treated correctly when it arrives.

Layer 5 The *session layer*: responsible for establishing and managing sessions between application programs. It will ensure that the remote user is connected to the correct application program for the purpose of the session.

Layer 6 The *presentation layer*: contains the functions of format and code conversion, making data more easily interpretable by the application program.

Layer 7 The *application layer*: provides the services, file transfer, electronic mail, etc. that the user or the application program requires.

And above all this, somewhere unspecified, hovers the human user for whose sake all this has been arranged.

35 Data Services and Packet Switching

1 Computer-based Data Links

The coming of the microprocessor means that many offices which could not economically justify the purchase of a large main-frame computer are now able to increase their efficiency by using digital computers to carry out many tasks which were previously expensive and labour intensive.

There is still however a huge demand for interworking between computers, minicomputers and microcomputers at different locations. The high-street banks, for example, make use of computers to maintain details of customers accounts, of standing orders, direct debits, etc., while airlines and package holiday firms are able to operate booking systems that provide an immediate confirmation of vacancies and bookings.

The cost of a digital computer with a large storage capacity is high and so it is not economic for an organization to install a computer at all the points in its offices and factories where a computing facility would be of use. It may, however, be economic for a main-frame computer to be installed only at one, or perhaps two, points in the organization's network of offices. For the branch offices to have on-line access to computing facilities it is necessary for them to be connected to the computer centre by means of data links.

Organizations sometimes find that it is attractive for various reasons to be able to share access to their computers and databases with other parts of their own local organization. To enable one computer to interwork satisfactorily with others in the same area, Local Area Networks, or LANS, have been developed (see section 37).

Many smaller businesses are unable to economically justify the cost of purchasing and operating a main-frame computer and yet have a need for computing facilities. To meet this demand computer bureaux have been set up which rent computer services to their customers on a time-sharing basis. Customers need data links to these bureaux.

2 Privately Leased Circuits and the PSTN

A data link that connects a data terminal to a remote digital computer may be leased from the telephone administration or it may be temporarily set up by dialling a connection via the public switched telephone network (PSTN). The choice between leasing a private circuit and using the PSTN must be made after the careful consideration of factors such as the cost, availability, speed of working, and transmission performance. Private circuits may transmit d.c. ± 6 V or ± 80 V signals or may use modems to convert the data signals into voice-frequency signals. DC can only be used if there is a metallic path all the way; if the circuit is provided by any form of multiplexing, either FDM or TDM, the signals have to be fed through a modem. Voice-frequency data circuits can also be of any length and may be routed, wholly or partly, over audio-frequency cable, or over multi-channel telephony systems. A modem (short for modulator—demodulator) converts digital signals from a data terminal into modulated analog signals which can be carried by the PSTN. These analog signals carrying digital information are sometimes called pseudo-digital signals.

In many countries the cost of permanently leasing a line may be relatively high and it could only be economically justified if there is sufficient data traffic on the line or the particular terminal application necessitates a permanent connection. If the data communication requirements involve occasional contact with a large number of locations, and the majority of the connections are of fairly short duration, the use of the PSTN is probably best. On the other hand if long-duration connections between a few branch offices and the computer centre are likely, leased links will probably be chosen. In practice, most private data networks consist of a combination of both leased and PSTN links, and very often the leased circuits are provided with the stand-by facility of using the PSTN when necessary (i.e. if the leased circuit should fail).

3 Switched Circuits

The use of the PSTN to carry data traffic often restricts working speed significantly, and call set-up time limits interactive working between terminals. In recent years many telecommunications administrations have recognized that there is a genuine public demand for improved data facilities, and are now beginning to provide public switched data networks so that customers are spared the expense of leasing long-distance circuits and avoid the operating constraints of using the PSTN.

There are three basic ways of switching record traffic or data:

a) Circuit switching: a complete communications circuit is established for the exclusive real-time use of the subscribers for the duration of the call.

b) Message switching: this is a store-and-forward technique which is used for message traffic but is not appropriate for all data purposes. The various types of magnetic or semi-conductor stores now used are a great improvement over the "tornpaper tape" stores which were commonly used in the 1970s. Even so, message switching can sometimes run into significant delays of minutes or even hours, making interactive traffic impossible. Since interactive working is normally essential between a remote terminal and a host computer, message switching is not practicable for such services.

c) Packet switching: data streams are split up into short packets with well-defined formats. Because there are no long data streams, the blocking of links is rare in well-engineered systems and end-to-end delays are small, often less than one second. Interactive working is practicable, as with circuit switching.

4 Packet Switching

In the ordinary circuit-switched Public Switched Telephone Network (PSTN) an end-to-end analog service is provided by setting up dedicated paths comprising fixed point-to-point circuits joined in temporary association by switched contacts at exchanges. Path set-up is controlled by the calling subscriber (see fig. 35.1). If we use a PSTN circuit to carry data, we need a modem (modulator plus demodulator) at each end of the

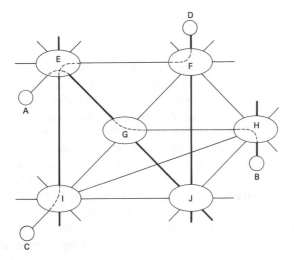

Fig. 35.1
Circuit switched network: calls from A to B and from C to D are established by paths interconnected at switching centres E,F,G,H,I

circuit to convert the digital signals of our data terminal to an analog form suitable for transmission over the PSTN. A good modem and a good circuit will permit a bit-rate of 4.8 kbit/s but on most circuits in most countries 2.4 kbit/s is all that can be managed without too many errors being introduced. It takes time to set up a PSTN call, often several seconds, and as the network gets busier the call-set-up time increases until there is congestion, and no calls can be established at all even if the wanted subscriber at the distant end is free and waiting for your call. Most telephone calls last for several minutes and in many countries you pay for calls in units, 1 unit call for every 3 minutes or less, so a few seconds for the call to be set up is acceptable. But data calls sometimes need last for only a fraction of a second and it would be very wasteful if you had to pay for 3 minute's use of a circuit which you had in fact used for, say, only one-tenth of a second!

Another way of switching data traffic is to use message switching, or "store and forward" switching (see fig. 35.2). Message switching is similar to circuit switching in that the network comprises a number of switches interconnected by point-to-point circuits, but, additionally, at each switching node there is a memory store into which messages may be passed for temporary storage or buffering. At each such node, the destination address of each message is examined to determine whether the message is to be transmitted on to another node or if it is to a destination terminated on that switch itself. If the outgoing link, or the circuit to the destination address, is free, the message is sent straight on. If the required link or destination is busy, the message is stored or buffered in a queue of messages in the local memory. Transmission of the message is tried again when it has worked its way again to the front of the queue; different priority levels can however be assigned in the queue.

Fig. 35.2

Message switching: store and forward: message from A to B is stored temporarily at C and E because circuits forward are busy. As soon as circuit forward is free, the message is pumped from the store on towards its destination

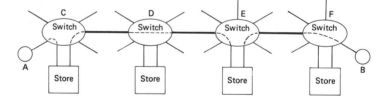

Message switching using store-and-forward techniques does therefore introduce a finite delay, which builds up as the buffer stores in each node fill with messages awaiting clearance. This means that message switching cannot be used for interactive communications, that is to say messages demanding immediate replies, because total queuing time can sometimes be minutes or even hours, not just nanoseconds.

Packet switching networks have been designed so that interactive data communication *can* be established. A packet switched network is a store-and-forward process as shown in fig. 35.2 but the messages are not of indefinite length, they are divided up into short fixed lengths known as packets. A few very long messages in an ordinary message switching system can hold long-distance circuits busy and make it impossible for other would-be users to get their messages transmitted. The division of traffic into small packets eliminates this difficulty. In a well designed system there is no blocking and queues are very small.

Each packet carries in its "header" (the first few digits) the address to which it is to be forwarded and, as with ordinary message switching, these "address to" digits are automatically analysed at each switching node and the packet either sent straight on its way or (if all onward circuits are busy or out-of-order) put into the local temporary store. This means that, on the main highways between switching nodes in the network, the packets received from one terminal are likely to be inter-leaved with packets from many other terminals. Because the "address to" digits are read and acted upon at each node, some packets may be sent one way and some on other routes to the same final destination. At the destination, the packets into which a complete message was divided are automatically reassembled into their correct order so that the message becomes a sensible whole.

A packet switching network can be used by "non-intelligent" data terminals like teletype machines as well as by specially designed packet-mode terminals. The signals received from a teletype or similar machine are dealt with by a PAD (Packet Assembler-Disassembler) which takes in signals character-by-character and builds them up into standard packets which it sends out to the packet network (see fig. 35.3). Because the network uses store-and-forward procedures it can be used by terminals using many different data rates. If a message from a fast terminal is to be delivered to a slow terminal, the message merely stays in buffer stores a bit longer, because the last link out to the destination terminal is restricted to the slower speed of that terminal.

Fig. 35.3
Packet switching network with packet assembler-disassembler (PAD) to interconnect non-intelligent teletypewriter to the network, showing packets interleaved on the link between nodes. PAD functions are detailed in CCITT Rec.X.3, and interworking with PADs is described in X.28 and X.29

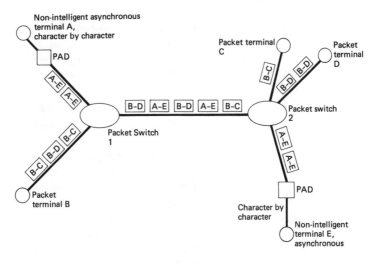

Each packet also has a trailer with an "error checksum", used for error correction. This error checksum ensures that the packet has been received without errors. Before the days of integrated circuits this procedure would probably have needed a whole mainframe computer of its own. The information in the packet, excluding the "flags" which indicate the beginning and ends of signals, is considered for this checking purpose as being a string of numbers, all in binary code. The total figure is divided by a 16-bit number (internationally standardized), the quotient is disregarded, and the remainder added to the packet to be transmitted as a separate checksum. At the receiving end, the whole packet including this extra checksum is again divided by the same 16-bit number. It should divide exactly, with nothing left over. If there is a remainder, it means something has gone wrong during transmission so the whole packet is retransmitted, automatically, until it *is* received correctly.

Packet switching is simple in its basic essentials as given above but extremely complicated in all its details, especially as these are all of international validity.

A few definitions will however be of interest:

A *datagram* is a packet which is a self-contained message; it is routed individually through the packet network rather like a telegram going through a message switching system.

A *virtual call* is a bit more difficult. If one terminal in a packet network wants to send a long message to another terminal, it begins by telling the distant-end node that a long message is coming and special indicating signals called logical channels are established. From then on all the many packets that make up the complete message are sent through the network just like other individual packets but at the destination end they are immediately assembled in the correct order and delivered to the destination terminal as one complete message. For all practical purposes, the two users have a direct private circuit from one terminal through to the other; the fact that the message has been divided up into hundreds of separate packets which might have followed different routes through the network to their destination is of no real worry or importance to the user, hence the name virtual call.

Protocol means the rules and regulations. For packet switching, the protocols are given in CCITT Recommendation X.25.

CCITT Recommendation X.25 defines the interface between a user's packet-mode terminal (Data Terminal Equipment, DTE) and the packet network whose entry point is called a Data Circuit-terminating Equipment, DCE. Figure 35.4 shows the relationship between DTE and DCE. X.25 also defines three basic levels, in line with the layers defined in OSI:

X.25 Level 1: provides a synchronous, bit-serial full-duplex point-to-point circuit for the transmission of data between user's equipment and the network (Rec X.21 defines the interfaces).

X.25 Level 2: is the link control or frame level, it deals with the detection of transmission errors and their correction. Another ISO specification is used here, that for High-level Data Link Control or HDLC, to provide the error-free transmission system.

X.25 Level 3: defines three basic types of packet service:
  a) a single packet service, the datagram
  b) a multi-packet service provided on a switched basis
  c) a multi-packet service provided on a permanent basis.

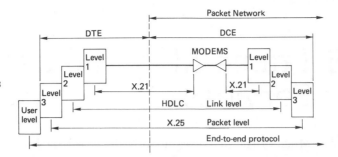

Fig. 35.4
A user's data terminal equipment (DTE) and its interface with the packet network whose entry point is the data circuit-terminating equipment (DCE)

International interworking between packet-switched data networks is now possible; there are now many such networks all over the world. Figure 35.5 shows how a terminal in one country can access a network in another country, using their International Gateway Exchanges, G on the diagram. An internationally agreed numbering plan has to be used; this is given in detail in CCITT Recommendation X.121, and is shown in fig. 35.6.

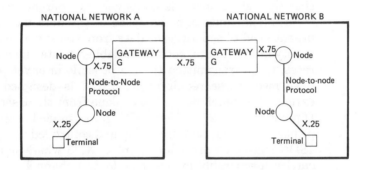

Fig. 35.5
International packet working using gateway exchange G, showing the protocols and specifications for each section

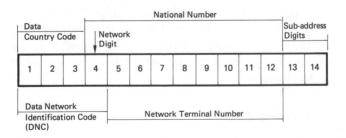

Fig. 35.6
The internationally agreed numbering plan for packet switched networks (CCITT Rec.X.121)

36 Frame Mode Bearer Services

Much of the protocols adopted for the ISDN use very similar methods to those pioneered in X.25.

Packet switching using X.25 is an example of an "enveloping" information transfer protocol (see section 27). The data to be transmitted is enclosed in successive envelopes of signalling information directed at the various layers in the information transfer hierarchy, the OSI model for example (see section 34). One disadvantage of the simpler forms of these protocols, such as packet switching, is that the signalling and the data are transmitted as one and subjected to the same error correction and retransmission mechanisms and the same flow control mechanisms so that the data transfer is no longer transparent from user to user.

As the amounts and the speed of the data increase, there is a more urgent need to separate the data from the signalling. This enables more bursty kinds of data, such as variable-bit-rate video and much higher bit rates, to be transmitted than is possible in enveloping protocols.

To meet these requirements ISDN is designed to separate, to an extent, the signalling on the C-plane from the User information on the U-plane (see fig. 27.3). New ISDN frame mode bearer services follow this principle. Virtual calls are set up and controlled using the same out-band ISDN signalling procedures, unlike X.25 where signalling and data are carried inseparably in the same logical channel.

Two distinct ISDN frame mode bearer services have been defined: frame relaying and frame switching. The same signalling procedures are used for both but they differ in the protocol supported by the network in the U-plane during information transfer. In either service the user is able to set up a number of virtual circuits simultaneously to different destinations.

1 Frame Relay

The simplest frame mode service, frame relaying, is shown in fig. 36.1 which should be contrasted with fig. 27.3. Signalling in the C-plane uses the normal ISDN signalling for layers 2 and 3 with some enhancement to support the frame mode parameters. In the user plane however the network supports only a part of the link layer (layer 1) protocol.

The basic service provided by frame relay is the unacknowledged transfer of frames between ISDN terminals.

2 Frame Switching

The protocol architecture for frame switching is shown, in the same format, in fig. 36.2. Again, signalling in the C-plane follows ISDN

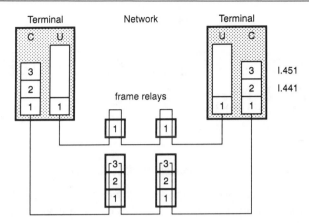

Fig. 36.1
Frame relay

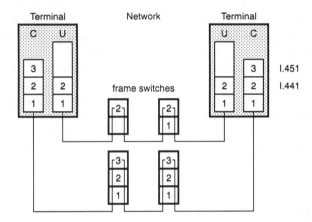

Fig. 36.2
Frame switching

practice but in this case the network operates the complete link layer protocol in the U-plane during information transfer. Frame switching therefore provides acknowledged transfer of frames between terminals. The network detects and recovers from lost or duplicated frames and frames containing errors and it will operate flow control.

The use of frame mode services considerably enhances the ability of the ISDN to carry a wider variety of applications. Allied to the broadband ISDN it may provide a vehicle for the transmission of voice and video services without the need for circuit mode connections. The lightweight frame relay protocol supports applications (such as voice) where additional delays cannot be tolerated. Frame switching offers the robustness against errors usually associated with X.25 but without the per-packet overhead of layer 3.

37 Local Area Networks (LANs)

1 Area Networks (LAN, WAN, MAN)

LANs have been defined as networks permitting the interconnection and intercommunication of a group of computers, primarily for the sharing of resources such as data storage devices and printers. LANs cover short distances (less than 1 km, usually), almost always within a single building complex. Different data transfer rates are possible and shared centralized data storage access is available. Response times are comparable with those of a single computer.

Networks which have been designed to carry data calls over long distances (many hundreds of kilometres) also exist; these are usually called WANs, or Wide Area Networks. WANs provide their long-distance facilities either by using modems (modulator plus demodulator) and the ordinary public switched telephone network or by using the digital transmission services which are now beginning to become available in many countries, providing paths at 56 kbit/s, 64 kbit/s, 1.5 Mbit/s, 2 Mbit/s or even higher bit rates.

One great advantage which digital transmission has over the use of modems results from the fact that modems convert digital signals from a data terminal or computer into modulated analog signals which can be carried over the ordinary telephone network. At the receiving end, the demodulation section of the modem changes the received analog signal back into digital form again. The limited bandwidth of analog circuits, the distortions due to analog/digital and digital/analog conversions, and the various noises which are picked up during analog transmission, mean that there are many sources of error and distortion, and severe bit-rate constraints arise with this type of transmission. Digital transmission eliminates these difficulties.

These are the main reasons why many users are now willing to pay more for the use of digital transmission systems for their long-distance links instead of using modems plus analog transmission.

A third type of data network is now beginning to be developed in major cities, with ranges between the purely local LAN and the long-distance WAN. These are called Metropolitan Area Networks, or MANs. MANs are likely to become common and extremely important as cable TV systems grow, and they could well develop into backbone networks used to carry digital transmission systems right across our cities at high bit rate and with minimum distortion.

Much of the communications facilities required for MAN and WAN installations will, of necessity, be obtained from the public network operators. There is no compelling reason, apart from cost, why all MAN and

WAN facilities should not be provided by the public network. Readers of section 26 will also have realized that the ISDN basic access is equivalent to a small LAN combined with a small PABX.

2 Transmission: Broadband or Baseband

There are two major methods of transmitting signals along the physical medium, whether fibre, coaxial cable or twisted pair wire, of the LAN: broadband or baseband.

Baseband signalling is the simpler of the two. Only one signal at a time can travel along the cable; changes in voltage represent the information. More than one device can use a baseband network provided that there is a method for controlling access to the cable.

Broadband signalling works the way radio and television work by splitting up the available frequencies into different channels. Broadband technology is more complex than baseband but a broadband network is capable of carrying more information and of supporting more devices.

Baseband signalling is suitable for most LANs although broadband signalling is used in large multi-building networks. The IBM PC Network was originally specified for use on broadband coaxial cable but baseband support was added later so that it could run on the twisted pair of the IBM cabling system.

3 Network Topologies

Networks can be of several topologies. The simplest is probably the *star topology*, where all devices are connected to a central switch. A data PABX would be of this type. It has the disadvantage that the entire network depends on the central switch.

A *ring topology* (fig. 37.1) permits fairly simple and reliable access techniques. Messages are carried round the ring visiting every device in

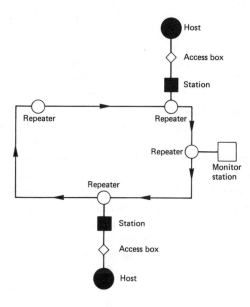

Fig. 37.1
A ring system

turn until the one to which they are addressed collects them. Again the reliability depends on all the devices in the ring.

A *bus topology* (fig. 37.2) does not depend on a closed ring but requires rather more sophisticated access methods to avoid all devices calling at once. Reliability is not as much of a problem.

Paradoxically, LANs are often realized as bus or ring but with the bus folded away in a central hub so that they appear to be working as a star topology.

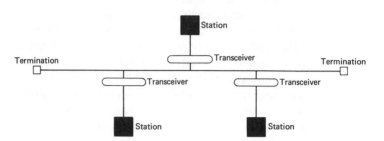

Fig. 37.2
A bus system

4 Access methods

Baseband signalling, the simpler of the two transmission options we have discussed, allows only a single communication on the network at a time. Clearly there must be a control mechanism since any device may try to send data at any time. The two most important access methods used in LANs are Carrier Sense Multiple Access with Collision Detection (CSMA/CD) and token access.

CSMA/CD, a method suitable for bus topologies, is rather like shouting down the corridor, where the listeners perform the CSMA part, listening, and the shouter performs the CD part, listening while shouting. If the shouter hears another shout at the same time he waits and shouts again. In more technical terms, under CSMA/CD a device listens to the network before transmitting, trying to sense the presence of a carrier indicating that the network is in use. Even while transmitting, however, the device continues to listen so that if another device is transmitting at the same time they both detect a collision. Having detected a collision the devices cease to transmit, wait for a period which is made different for each device on the LAN, and try again. CSMA/CD is the access method used by Ethernet, perhaps the most widely used of all the LAN protocols.

Token access methods are more suitable for ring topologies and are a little like the Circle Line on the London Underground. Information packets are continuously circulating around the ring and each device can load information onto a packet if it sees that the packet is empty. Having loaded the information the device, or another supervisory device, can watch to check that the packet returns empty, indicating that the recipient device has received the message by unloading it from the ring. The byte of information in the header of the packet is called the token and is altered to busy by the sender when loading the packet and altered to free by the receiver when the packet information is unloaded.

A particular version of the token access method is token passing, much like single line working on the railway. In order to transmit, a device must seize and keep the token and can transmit for as long as it retains the token. When no longer transmitting, the device replaces the token on the continuously circulating information on the ring so that another device wishing to transmit may pick it up.

The IBM Token-Ring Network is the major example of a token-passing LAN.

Token access is an important concept for another reason: it has been chosen as the access method to be used by Fibre Distributed Data Interface (FDDI) LANs.

5 Fibre Distributed Data Interface (FDDI)

Ethernet, for example, operates at a speed of 10 kbit/s but, because of overheads and delays, its information transfer rate in practice is considerably less than this. If an Ethernet LAN grows to a size of more than about 80 devices then special design methods must be used to provide users with an acceptable speed of operation. If, working on your stand-alone personal computer, you can call up files almost instantaneously, you won't be best pleased if, on the LAN, you must wait 30 seconds to 2 minutes for your file.

Increasingly also it is becoming economical to wire up the work place with optical fibre. There is need for a LAN protocol that will utilize the much greater bandwidth available with fibre.

FDDI operates at 100 Mbit/s and there are developments proceeding to upgrade this speed to 200 Mbit/s. It is a ring system but with much greater security than the low-speed LANs. Information is transmitted by all devices at 100 Mbit/s in both directions around the dual, contra-rotating ring. If any device or any link fails then the neighbouring devices loop back the information so that the remainder of the ring still operates. (See fig. 37.3.)

An FDDI fibre LAN could be used as the backbone LAN serving Ethernet or other low-speed LAN branches. The FDDI devices would be routers, bridges or gateways interconnecting the FDDI LAN with the departmental LANs running at lower speeds.

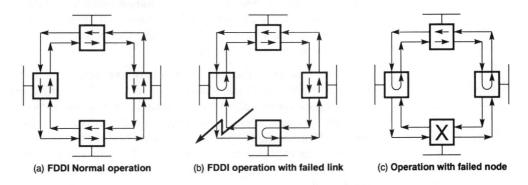

(a) **FDDI Normal operation**   (b) **FDDI operation with failed link**   (c) **Operation with failed node**

Fig. 37.3
Fibre Distributed Data Interchange local area network

Part G
Business Services

38 PABXs, Keyphones

When telephones were new and rare, very few companies had more than one line. The phone itself was carefully guarded and used only by The Boss. As time went on, the convenience of being able to talk to others without leaving your own desk encouraged companies to install phones for more and more of their staff—but here the first difficulty arose: cost. Each phone had to be complete with its own pair of wires all the way back to the central office/exchange, rented from the Telephone Company, and in many cases there was a charge for each call even when The Boss spoke only to an assistant in the next room.

There are basically two ways of tackling this problem:

a) Provide an "isolated" local telephone system in the office, an "intercom" with no connection at all to the public network. An automatic version of this is usually called a Private Automatic Exchange, PAX. Only a few senior people then have phones on outside lines.

b) Provide a switching system which enables the same telephone to be used for both internal and external calls, a Private Branch Exchange or PBX. Since the telephones themselves are not directly connected to an exchange/central office, it is usual for all the phones on such installations to be called "extensions". Early PBXs were looked after by operators; typically one girl could readily control a board serving up to about 50 extensions and 10 exchange lines. The operator dealt with all calls, internal and external, answered all incoming calls, and dialled out all the outgoing calls. Boards like this were called PMBXs, Private Manual Branch Exchanges.

This was fine; the operator knew where everyone was and it provided a built-in security service which stopped the office junior making expensive long-distance calls to his relations in Kansas City—the early and completely non-automatic ancestor of today's complex Telephone Information Management Systems (TIMS). But sometimes, operators being human, unacceptable delays occurred: if the operator simply had to leave her board for a while, neither outgoing nor incoming calls could be connected.

The development of automatic switching equipment soon led to the situation in which each instrument was able to control its own outgoing calls—it being about as easy to dial a number as to repeat it verbally to an operator—but even with these Private Automatic Branch Exchanges (PABXs), incoming calls depended on being answered by the operator. There is no easy way an ordinary outside call can be steered through a PABX to a particular extension.

Many modern public telephone central offices/exchanges apply a "time-out" to calls which are not answered within a specified period, usually one minute. This releases all the switches in order to stop the whole network becoming congested by connections which carry no traffic—and earn no money for the telephone company. If PABX operators are very busy (or if some of them are on unexpected sick leave), then by the time they get round to answering the last of the calling lines there is sometimes no-one there. It is not always the caller who has hung up in disgust, it is sometimes the switching equipment which has "timed-out" the call. This is probably one of the most frustrating aspects of trying to call busy companies: inadequate provision of operators must lose the companies concerned a lot of goodwill, and a lot of business.

One of the ways in which this particular difficulty can sometimes be overcome with modern switching equipment is to use central offices/exchanges with "in-dialling" or "dialling-in" lines. These are specially designed to enable outside callers to steer their own calls through the destination PABX. Another bright idea (which does not necessitate any changes in the central office or main exchange) is the use by the customer not of a centralized local switching system (a PABX) but of a system with very little centralized brains. Intelligence is distributed out to the telephones themselves; they have to be able to do more than just dial out digits. This was the origin of the keyphone system. Instead of incoming calls going only to the operator to answer, all the exchange lines were fed to all the extension instruments, giving all phones complete access to all lines. Any extension can answer incoming calls, all extensions can pick up outgoing lines and dial their own calls; no separate operator's position is needed, just an agreement by which nominated extensions answer calls and transfer them to the wanted extension.

This seems so simple that there has to be a snag, and there was: early versions of this sort of system needed large cables with a great many pairs of wires to each phone. If people's desks never had to be moved, this would not matter a great deal but in most active companies the reorganization of local layouts seems to be a frequent necessity as new departments or management structures are created, and desks have to be fitted into available space. Fortunately the chip came to the rescue here; it is now possible to purchase keyphone systems which do not need large cables, so it might be thought that it could now be a straight commercial fight between (a) systems using PABXs (i.e. a centralized switch) and (b) those in which much of the intelligence has been distributed to somewhat more complex keyphones. This is indeed the position today in many national administrations, the customer can choose either *a* or *b*.

More history now. When Common Control of central offices (mainly of crossbar switches) became economically possible in the 1950s, it was quickly realized that there was no purely technical reason for customers to have their own PABXs at all. Each of their extension phones can instead be served from the main switch or exchange; each is then given a standard number so incoming calls go straight to the required extension, with an automatic transfer to an assistance operator in the customer's premises if the extension fails to answer. Calls to the company's directory number also go straight to the assistance operator's console (often found at the receptionist's desk). The central office/exchange is programmed so that calls from one extension to another in the same company are ignored by the charging equipment, which would normally count each outgoing call as a unit call to be charged and paid for in due course. There is no PABX to purchase or to worry about or to maintain, it sounds too good to be true.

Cost is the main problem of course. A separate cable pair has to be rented from the telephone company linking each extension with the main central office/exchange and of course one complete line termination on the switch has to be rented also. Here a lot depends on the tariff policy being followed by the telephone company. If the central office/exchange is very close, the cable pairs may not be expensive. This service, using TelCo's central office as your own PABX, is usually called a Centrex service. There are variants of this available in some areas by which more of the switching equipment is located in the customer's premises; this reduces the number of cable pairs to be rented.

In some countries there is now yet another variant: Centrex-type service provided by a third party (e.g. the building landlord). When you rent an office in a block with this service, your office comes complete with mains power, water, sewage and telephone service: all the utility services are provided by the landlord. Ordinary PABXs are sometimes also installed in this way, one very large PABX in the basement serving each floor of a high rise building with different directory number groups available for the companies occupying each floor. Station message detail recording (SMDR) can be provided as part of all modern PABX installations. This enables the responsible manager to see where costs are being incurred and to take whatever action is appropriate. Telephone Information Management Systems (TIMS) have come a long way from their Kansas City origins!

Until about 1983 PABXs everywhere were analog devices, using space division switching. The adoption of digital technology (which in most countries began with large trunk and tandem switching centres) has led quickly to the digitalization of PABXs. It is now possible in many areas to install digital units which are not only cheaper than their analog counterparts but are able to switch both voice and non-voice signals. A single PABX can sometimes now be used as the controlling centre for a complete Local Network, able to establish calls between telephones, word processors, teletex machines and computers, all operating at different bit rates. Those multi-purpose switches have made complete Office Automation available now, not just a dream for the future.

If your business involves a great many incoming calls, any of which could be dealt with by any of a number of your people, then you will not really need a PABX at all; you need an Automatic Call Distribution system (ACD). Many firms now make these, indeed they may be supplied as a part of a PABX installation for just some departments of the firm to use. They can be associated with a computer terminal or VDU so that, while each caller makes his requirements known, the operator is keying in the particulars and getting all the relevant information displayed on the VDU screen to study and act upon. First-time callers always seem most impressed when, having merely quoted a policy number or a file reference, the operator calls him by his correct name and asks if he still lives at 5 Market Street.

ACD systems such as these have played an important part in the telephone share buying campaigns in the UK as each successive nationalized industry has been sold off to the public.

Brokers and Dealers have very special requirements which often involve another specialist form of PABX known as a Dealer Board. From each broker's desk a keyboard allows speedy access to 2, 4 even 6 hundred lines worldwide. The broker can hold several calls open at once and may wish to speak and listen on several open calls at the same time.

So if you need telephones for a new office you must find out first exactly what is locally approved and available. In many areas it is now possible to get genuinely competitive bids for the provision of identical services by quite different methods. Most modern systems are able to provide service features which were expensive luxuries—or completely unavailable—only a dozen years ago.

39 Virtual Private Networks, Centrex and SMDS

Modern networks incorporating intelligent network capabilities and ISDN and BISDN access possibilities can allow the user more ability to manage the network, both the user's own private network and the user's access to public networks and services. The intelligent network and the introduction of ISDN have also brought back into prominence the concepts of Centrex and Virtual Private Network (VPN).

At approximately the same time the marketing departments of the network provider have found new ways of selling existing or emerging products so that this section deals also with the subject of Switched Multi-megabit Data Services (SMDS).

Centrex is the provision of private communications switching function by means of public facilities located in the public telecommunications network. The Centrex service will be more conveniently provided by the local exchange as shown in fig. 39.1 but in the early stages of implementation it can be resident in a special exchange remote from the user. This latter arrangement is that used in the initial Mercury 2110 service in the UK and in that originally proposed, but not offered, by BT.

It is not necessary to talk of virtual networks in terms of Centrex but

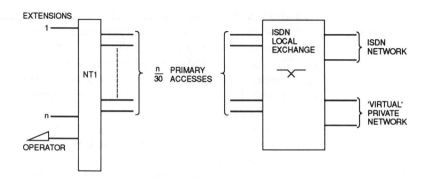

Fig. 39.1
ISDN Centrex service

it is certainly easier to do so. In switching, particularly digital switching, it has become convenient to refer to a permanently or semi-permanently held "connection" through the switch as a virtual circuit. Such a connection could be a permanent path between a terminal and a register or such-like device.

When the user rents a private circuit from the operating company, what is rented is the assurance of permanent connection between two terminals. The user certainly does not rent a particular combination of lines and channels; the only parts of the connection which are permanently defined are likely to be the local ends. Between these the operator can and does vary the connection path. The operator may however have to declare to the user certain characteristics of the line and therefore can only vary the routing of the line within the declared parameters. To this extent, therefore, every private circuit is virtual. In digital working, of course, the connection is re-established for every time slot and hence the term "virtual" came into use to describe a virtual permanency.

For a virtual private circuit the proposal is to market a technological fact and offer not a permanent guarantee of connection terminal to terminal but an assurance of connection when this is required. The user performs the same actions as would be necessary to seize a private circuit but the connection is made on request out of the full stock of options available to the public network.

To the same extent that Centrex provides a virtual private switch, the virtual private network interconnects the Centrex "nodes" so that the user finds the difference between the virtual and the real indistinguishable. Clearly there can be a real and significant difference in that the virtual approach can be far more economic and tariffed accordingly.

Provision of Centrex and virtual private networking leads to Virtual Network Management. Perhaps the greatest good to the user to be expected from the combination of Centrex and virtual networks is the possibility for the user to arrange the private network from a terminal.

Just as Centrex allows a private operator to deal with staff and the public as if this were a private exchange, the extension to a virtual network allows the user to effect changes electronically. Within organizatonal limits, set by the network operator, the user can add and change facilities, terminals, private links, etc. and can obtain full and immediate details of the effects, in costs and traffic, of such changes.

One example of a virtual private network is illustrated in fig. 39.2. The

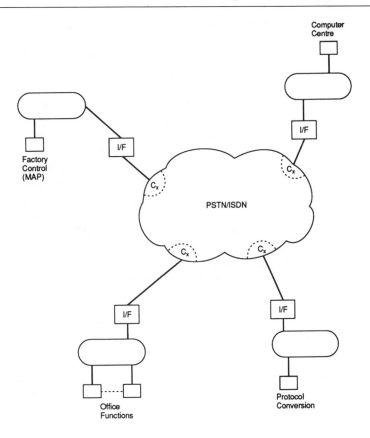

Fig. 39.2
Virtual private network
for data

network includes automated factory locations, office locations using a considerable amount of IT and a mainframe computer serving the industrial group. In such an arrangement there may well be interfacing problems in feeding information from (say) production to office terminal systems or to the mainframe data banks. The presence of CCS from terminal to terminal (CCITT Q.SIG or, in the UK, DPNSS) makes it possible for the terminal to define the virtual connection. Such a connection could, if necessary, include protocol conversion features provided as an external network service, like those shown in fig. 39.2, or as an internal service of the organization.

The combination of the ISDN basic access, representing, as it does, a small combined PABX and LAN, with the concept of the virtual private network is very powerful indeed. Organizations could operate as if there was but one facility interconnecting all their main locations and their high street branches without the need, in most cases, of setting up their own private network.

All that has been said may well be called into question by recent European Community Directives seeking to impose Open Network Provision (ONP) on the public networks within the EEC. The objective is to ensure that all public network operators offer a basic minimum of typical services. One of these that differs widely within the Community at present is the provision of leased lines. Section 40 covers these issues in more detail. What needs to be said here is that the European standards now being introduced to mandate a common minimum provision of ONP Leased Lines throughout Europe ought to take account of the possibility of the leased line provision being made through the means of virtual private networks. At present there is no real assurance that this will be the case.

Telecommunications has not been noted for sparing the public from the necessity of learning new words, most of them made up of initials. Switched Multi-megabit Data Services (SMDS) is one that does not represent a new technology so much as a marketing man's bright idea for selling something already in stock. Having successfully offered to the public 64 kbit/s and 2.048 Mbit/s links there is still a need for yet higher capacities for the large volumes of data required by some major private network users. Even so it is difficult to sell this capacity in the digital hierarchy units of 8 and 32 Mbit/s. SMDS is an attempt to provide the user with the equivalent capacity by means of digital connections not unlike those envisaged in the virtual private network.

Having started from a concept in the USA of providing switched data

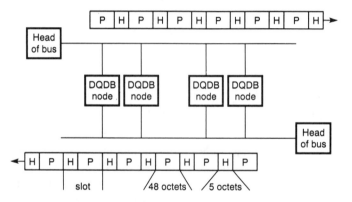

a) **Distributed Queue Dual Bus subnetwork**

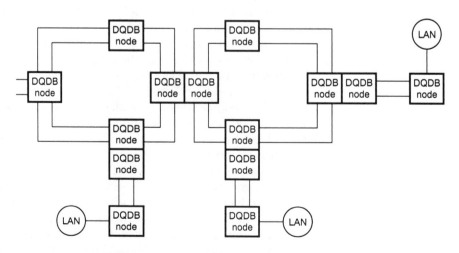

Fig. 39.3
Distributed Queue Dual Bus architecture and networks

b) **DQDB in a public data network**

services on demand and in the capacity demanded, SMDS has developed into a concept of a fast packet switching service not unlike frame relay (section 36) but with built-in security and very similar indeed to ATM (section 29). This maturing of the concept has brought with it a real technological advance in the form of the protocol most often envisaged with SMDS: the protocol, or interface technology, known as Distributed Queue Dual Bus (DQDB).

Each terminal on a DQDB bus (fig. 39.3) is required to count empty packets so that enough packets for information waiting further down the bus get to the end of the bus before the terminal seizes a packet for its own use. The packets are identical to those used in ATM (section 29). Whole networks can be constructed from DQDB buses as is illustrated in fig. 39.3.

Trial SMDS services are widespread in the USA and have been announced in the UK. The prospect of SMDS being freely available in the near future has called into question the advisability of choosing options such as frame relay and even perhaps endangered the introduction of VPN which is at present only available on a limited scale for international connections.

40 Private Networks and Leased Lines

The previous section suggests that private circuits and networks could become a thing of the past when the public network is able to provide the same, or better, functionality at a possibly reduced cost. Private networks have however grown in a climate of vastly diverse costs and considerable sentiment.

Table 40.1 shows typical costs for leased lines in several European countries. The tremendous differences explain why there are far more and far bigger private networks in the UK than in Germany, for example.

The differences in costs reflect a difference in attitude to leased line provision by the network provider. If the network is well provided with transmission links, and if the costs of calls over the public network are set at a rate that yields only a small profit, then the network provider will be keen to lease some of the network for an assured, profitable return regardless of whether it carries traffic or not. By contrast, providers whose networks are not so extensive and whose call costs are profitable will not be keen to lease lines except at quite high rentals.

In Germany, for example, leased lines are rented for the equivalent of the call charges for 80 hours use of the public network per month, a realistic but high figure that discourages the renting of circuits. In Germany therefore, private networks are largely limited to short

	France	Germany	Italy	Netherlands	Spain	Sweden	UK
64 kbit/s							
Connection	414	418	282	1319	1856	1189	900
Monthly, 50 km	1392	1842	1482	959	2258	317	316
Analog 2w M.1040							
Connection	207	209	188	309	221	827	1175
Monthly, 50 km	318	453	493	198	484	115	130
Conversion rate	9.67	2.87	2128.57	3.28	181.06	10.51	1

Table 40.1   European leased line rentals [with acknowledgements to Eurodata]

distances within the urban area and all longer-distance voice communication is carried by the public network. Data links and private networks do exist but the German communications manager will make very certain that they are well utilized.

In countries with low leased line rentals, such as the UK, particularly in the bad old days of a monopoly public network provider, there was every encouragement for the communications managers to escape the inefficiencies of the public network by setting up their own private network. Nowadays, with far more efficient public networks the private networks still proliferate in the UK but with far less reason.

The other causative factor that has encouraged private network growth has been the need for computer networking. This has been covered, in its technical aspect, in section 37. While transmission was analog there was a need to interconnect the computer entities with dedicated lines. The only lines available were those designed for voice transmission so that modem manufacturers designed their products to work over ordinary voice lines. Clearly each line could only carry communication between a pair of "user" computer devices. As data transmission was required at higher speeds, then special quality leased lines were engineered and they should be used above, say, 4.8 kbit/s. Nevertheless, many computer networks operate satisfactorily at 9.6 kbit/s and above over ordinary quality lines but the manager has little or nothing to complain about to the leased line provider if they fail to operate satisfactorily.

Today, we have available for rental, from the network providers, ordinary and special quality analog lines and digital connections operating at 64 kbit/s or 2 Mbit/s. The digital lines can be used for several multiplexed data or voice connections.

The modern network manager can therefore engineer the private voice network and the private data network using, quite often, the same leased line facilities. Whether they should continue to do so is quite another matter.

PABXs equipped with least cost routing algorithms, as many of them are today, could often achieve the same results in maximum economy in call handling without any leased line connections in the network at all. The leased line rental is payable regardless of whether it carries traffic or not. It could still be economical to use leased lines for data where the computers are able to utilize the line 24 hours a day or where the line can be used for voice during the day and for data overnight.

All things being equal then, with the advent of the ISDN and using Centrex and VPN, we might expect to see the use of leased private networks decline. Two developments make this unlikely.

The first affects only the UK at present. The licensing arrangements have recently been relaxed as a result of the Duopoly Review of BT and Mercury conducted in the latter half of 1991. It is now possible for many more organizations to offer public network service. First into the fray were organizations, like the Post Office, British Rail, even British Waterways, with actual or potentially over-provided private networks.

In such a situation one would expect the existing public network providers to be rushing to offer Centrex and VPN at most attractive rates in order to ensure that fewer private networks had excess capacity to sell on. However, as one German telecommunications man put it: "When Telekom suggests Centrex then all the PABX manufacturers will put on their dark suits and go to Bonn." There are unwritten understandings in many of the countries in Europe that Centrex will not be offered aggressively lest it damage the PABX industry.

The second development comes from the EEC. The Commission sees it as a duty to ensure that the same commodity is available on the same terms throughout the Community. We have become used to speaking of a "level playing field". In telecommunications this endeavour has been expressed in the Open Network Provision (ONP) Directives. One of these concerns leased lines and the European Telecommunications Standard Institute (ETSI) is engaged at present in specifying the set of ONP leased lines which must by law be available at equivalent charges anywhere within the European Community. It is not yet clear whether this minimum set of equally available leased lines will include virtual circuits. This development could well change the perceptions of the communications manager, particularly those in countries where leased lines are not an economical consideration at present.

This is another of those cautionary tales where technical and economic developments interfere with each other and magnify the difficulties of rational technical choices.

41 Call Logging

Call logging is a peculiarly British solution to a problem that is preponderantly a British problem. In Strowger step-by-step telephone exchanges, the charging information consists of periodic pulses sent back over the network to step on an electro-mechanical meter located at the subscriber's local exchange.

Even today, many of the lines in the UK are still equipped with such devices which are read by being photographed and the photographic record input manually to the computer systems which produce the subscriber's bill. In fact, the Post Office (as it then was, British Telecom nowadays) was so satisfied with the system that it refused to allow the manufacturers of the first stored program control and, later, digital switching systems to include facilities for automated call charging.

Thus, until quite recently, British Telecom has absolutely refused to consider the provision of detailed accounts for telephone charges. The subscriber has had to be content with a bill which indicated the line rental and a total for all metered calls.

Mercury, when it started, used modern exchanges and provided detailed billing as a matter of course and BT, urged on by the Regulator OFTEL, is now offering detailed billing as a free option in those exchange areas which have the facilities to do so.

In the absence of detailed billing there was an urgent need for communications managers to be able to see where they were spending their money on telephone calls. Call logging equipment has been available from the early 1970s for installation as an addition to the PABX; and modern PABXs, in all but the smallest sizes, offer call logging facilities as part of the installation.

Provision of call logging is essential to those organizations such as solicitors or consultants who charge their communications costs to the client. They are highly desirable for any organization which wishes to control departmental expenditure in this as in any other area. The authors know of companies who announced the installation of a call logger and experienced significant reductions in telephone expenditures well before the equipment was brought into service. Companies are well advised never to announce when the call logger is removed and not replaced.

The detailed bill from Mercury is probably good enough to replace the need for a call logger. Mercury will itemise the bill down to departmental codes provided by the customer. The present British Telecom facilities are not so extensive. It may be therefore that the provision of a call logger cannot any longer be justified for cost control purposes. The facility is still necessary however as a means of keeping an eye on the performance of the PABX and the private network.

If detailed traffic analysis facilities are not available as part of the PABX then a call logger will have to be used to obtain them, particularly when extensions to the network or replacement of switches is being planned. Modern PABX networks, permitted by the regulations to carry traffic destined for the public network, must show that they provide the necessary minimum grade of service to the public network traffic. This is more than a detailed bill, even from Mercury, is able to provide.

42 Telex

Telex had its origins in the telegram services. In the 1930s large companies which needed to send and receive many urgent telegram messages between their branches persuaded local telecoms companies to rent point-to-point teleprinter circuits to them. They used almost exactly the same machines and transmission circuits as were used for the public telegram service except that the messages were printed out on ordinary paper instead of on the thin paper-tape which was then used by Post Offices for public telegrams (GPO operators stuck the received paper-tape down on to telegram forms for delivery to the addressee). Later on it became generally accepted that such messages could, despite the absence of an authenticating signature, be regarded as genuinely authentic even outside the purely in-house range of a privately rented circuit and a switched service was provided. To reinforce the legal status of the message the machines used with the switched service were designed to send an Answer Back message when they were called, identifying the receiving machine. Thus a telex message is authenticated by the answer back code recorded at the sending machine.

In most countries the switched telex service started off by being manually controlled by operators, just as the trunk telephone service was in those days. But as with the telephone service, traffic grew too rapidly to permit the handling of calls by human operators to continue. Early automatic telex exchanges were very similar to the automatic telephone exchanges of the same era, you had to dial the number you wanted. Telex exchanges in many countries tended however to become computer controlled somewhat before local telephone exchanges, mainly because telex signals were already carried by digital codes which lent themselves to computerized handling far more readily than the complex analog voice signals encountered in telephony.

In recent years there have been many improvements made to telex services; modern telex machines are just as quiet in operation as modern typewriters, and operators are able to prepare and compose their messages on TV-type Visual Display Units. Only when satisfied that there are no errors in the copy is the message sent forward to the system for dispatch to the addressee.

Another reason for the continued popularity of telex is that you do not now have to have a dedicated telex machine—an ordinary personal computer with the right software package and a modem can be used to send and receive telex messages.

It is anyone's guess at the present time whether or not the ISDN will one day get people used to the idea of data flowing speedily between offices, or whether the humble telex will continue to live on, despite it being able to operate only as fast as a competent typist, say at about 50 words per minute, whereas modern communicating wordprocessors and fast printers can deal with traffic at over a hundred times that speed. It will probably be many years before all our telex services can be pensioned off.

43 Least Cost Routing (LCR)

Least cost routing is a procedure by which outgoing telephone calls are routed out over those circuits which cost least. It is therefore only practicable when there is a choice of carrier or of tariffs.

A simple example will suffice. If your PABX has a direct tie-line (leased from BT or Mercury) to your factory in a distant town, then it should always be cheaper to use this tie-line rather than to route calls out on the ordinary dial-up public network. If the tie-line happens to be busy when you initiate your call, an LCR-equipped PABX can be programmed in various ways to deal with your call request, for example:

a) to route your call immediately out over the public network despite the extra cost

b) to keep monitoring the situation and if the tie-line is still busy in say 5 minutes time then to route you out on the public network, or

c) to make you wait until the tie-line does become free and then to put through all waiting calls, in queued-up order.

When there is more than one carrier an LCR-equipped PABX will choose the one which is cheapest for you at that particular time. The time-of-day and the date are of special importance in the USA where what is called a WATS tariff can apply. Under these Wide Area Telephone Service arrangements you pay a higher rent for your line but this gives you a specified number of "free" minutes of long-distance calls per month to specific zones of the country. If you have not used up all your allocation of time for calls to Los Angeles then a call out to Chicago could well be routed out via Los Angeles on the line for which you have paid but not fully utilized. So LCR equipment is not quite so simple as it might be thought. We do not at present have any WATS-type tariffs here in Britain but they might well come in. Here we only really have to choose between leased tie-lines, BT and Mercury.

In the UK, therefore, the nearest common equivalent to the USA's LCR units are Mercury's Smart Boxes. With one of these on your PABX your local calls go out via BT as usual but all long-distance calls are routed out via Mercury. All very simple compared with LCR in the States, where these units often have 7 or 8 possible outgoing routes to choose between, many of them with tariffs directly affected by cumulative total traffic carried on each route concerned.

44 Direct Dialling In (DDI)

When private exchanges were first introduced, a compelling reason for doing so was to have a uniform, polite, efficient answering service for the organization's incoming telephone calls. Forgetting this reasoning, there has been increasing demand in recent years for the facility to bypass the PABX operator and dial in direct to the PABX extension. It is worth stressing at the outset that DDI is only of value when all the other facilities of the modern PABX or Centrex, such as call diversion on busy/no reply, are available *and used*.

Public network operators have been reluctant in the past to provide DDI except in rare cases or for a rather high price, because it uses up scarce resources in public network numbering (see section 54). This is not as true of modern switching systems where the DDI number can be made an extension to the full public number but was certainly true of step-by-step systems where the DDI number subtracted from the capacity of the local exchange.

A few years ago, when BT was still part of the British Post Office, they had to fight very hard for all the funds needed for their investment programmes and probably could never get enough to be able to afford to take whole groups of lines out of the normal allocation lists—but they did manage to do just this for their own Head Office. Forty years or more ago direct dialling-in was given to BT's own staff while they said it just could not be done for ordinary mortals!

With modern computer-controlled exchanges there are no technical difficulties in providing DDI for a proportion of lines on each exchange.

While DDI is one of the ISDN Supplementary Services (see section 46), it is possible that the demand for the service may decrease with the wider availability of feature-rich telephones and ISDN facilities. Certainly it is a mixed blessing if one is connected directly to the telephone in an empty office and the office holder has failed to either divert the telephone or switch on the answering service.

DDI is provided as a facility over special public lines to the PABX; the ordinary exchange lines are still accessed using the single PABX directory number and are answered by the PABX operator. There is an analogous service to DDI, originally called, confusingly, Direct Inward Dialling (DID) but now more properly referred to as Direct Inward Access. This facility is quite often used in conjunction with facilities such as hunt groups or call distribution systems within the PABX. For Direct Inward Access a group of ordinary exchange lines are given a different directory number to the main PABX number and calls to these lines are routed straight to the hunt group or call distribution system. This is a very valuable, and cheaper, alternative to DDI.

45 Conference Calling

In a "traditional" analog network there are transmission complications whenever you connect more than one telephone instrument to the same line. If the incoming call is a bit faint to begin with and the incoming signal is shared out between several ordinary phones at the listening end, the net result can easily be that the caller's voice becomes so faint that no-one can hear satisfactorily. One solution was for a small extra earpiece to be attached to your line and a colleague would be summoned to your office to hold this high-impedance low-loss earpiece to his own ear to hear the distant end's responses.

With the invention of the transistor, "conference bridges" became available. These comparatively complex devices came complete with amplifiers and your PABX operator could arrange for several of your office executives to be connected together via these bridges so that a group of users could talk together to each other and to the distant end.

An alternative was to set up a meeting with several people gathered around a standard loud speaking telephone. Difficulties were often encountered, many of them due to the fact that the design of loud speaking phones has to be such that feedback between its loudspeaker output and its microphone is not permitted, to prevent the system "singing". There is usually some form of muting provided so that, when you are speaking, your own loudspeaker circuit is turned right down. This means that if the other end tries to butt-in and disagree with you, you cannot hear a word they say. That is difficulty no. 1.

Then if two of your colleagues start whispering confidential arguments to each other you can have no assurance that these whispers have not been loud enough to activate the microphone at your end, giving the distant end the benefit of all these in-house arguments.

With digital transmission and switching the conference bridge becomes a rather easier device to realize. It is, basically, an adding circuit that adds together all the contributions to the conference except your own, thus getting the effect of the muting loud speaking telephone without the disadvantage of cutting off potential interrupters.

Thus a user who is allowed conference facilities may put together a conference of, usually, up to 20 people located within the same PABX or private network, within the same national network or anywhere worldwide.

Conference facilities between two or more meetings, where many of the participants are in the same room, is a little different. For this one needs a microphone system in the conference room of a suitably high quality. Such arrangements are known as teleconferencing.

The addition of pictures on screen adds another dimension. Travelling around for meetings has become ever more tedious and expensive. The provision of pictures of the people at the other end of a telecoms line has become cheaper and easier, largely because it is recognized that a high "broadcast quality" moving picture is not necessary for business conferencing purposes.

Complex algorithms which separate out moving images from static background enable high-quality video to be provided for conferencing from a relatively narrow-band transmitted signal. This is usually called "full motion video". At a slightly lower level various techniques have been adopted by different manufacturers to produce adequate-quality still pictures, refreshed at intervals of a few seconds. These systems are usually called "near motion video". Some of these systems will even operate over an ordinary voice-grade circuit, something that was quite unthinkable 10 years ago. Other systems use wider bands. In general you get what you pay for: the wider the frequency band the better the quality of the picture.

People in the developed world are, however, still resistive to the idea of going across town to a purpose-built teleconferencing studio to talk to colleagues who have had to travel to a similar studio at their end also. There are many video-conferencing systems now in successful use but almost all of these are operated by large companies for their own internal management purposes. They find that executives do not mind going to a special conference room in the same building complex to talk to colleagues who have gone to a similar room at their end. If papers or the attendance of individual specialists are needed they can always be brought rapidly up to the conference room from nearby offices. The Boeing Group, for example, has now had special video-conferencing studios built at most of its many plants in the USA. Originally they expected a 15% reduction in staff travelling and hotel costs which would have paid off the costs of their first custom-built studio in one year, but the actual savings proved sufficient to pay it off in only 5 months, so Boeing decided to go into video-conferencing in a major way. They say it cuts many months off the design and manufacturing periods for major aircraft.

The use of common-user specially equipped conference rooms, which could be linked by special circuits to other similar rooms in other cities, is still however in some countries a thoroughly practical and viable project especially if actual physical travel to some of the cities concerned presents problems.

There is still some resistance to the idea of making electronic signals travel rather than people, by the very people directly concerned.

Does the physical presence of people produce so many advantages to the company that the extra costs associated with actual travel should be accepted? Long journeys for many people to attend a single meeting can only rarely be justified, so although common-user teleconferencing facilities may not yet make much profit they will in the next few years become an accepted method of doing business at a distance.

With the ready availability of very wide bandwidths in optic fibres, one wonders how many years it will be before holography can be brought fully into teleconferencing also. How useful it would be for some (but how horrifying for most of us) if a life-sized, fully three-dimensional and completely solid-looking hologram image of the Big Boss could be produced to lecture to gatherings held simultaneously all over the continent.

46 ISDN Supplementary Services

The previous sections have been discussing particular examples of supplementary services, services additional to ordinary telephony which the network or service provider may wish to provide, possibly for an additional charge.

Many such services become rather easy to provide when common channel signalling allied with digital transmission is extended right out to the subscriber's terminal. This was the proudest and most exciting claim for the ISDN. The very fact that we have been discussing some of these services, existing in the network without ISDN, indicates one of the problems of introducing the ISDN to provide services that users have already been given by other means.

Table 46.1 indicates the richness of the supplementary services that will become available as the ISDN is introduced. Against each service is the identity of the CCITT Recommendation already existing or in process of being drafted to describe the supplementary service. This is a simplified view; in their wisdom the standards makers have drafted three documents for each service. Table 46.1 shows only the most detailed level of these three standards.

It is the scale of the opportunities with ISDN supplementary services which is so impressive. We have become accustomed to transferring our telephones around the office as we go to meetings or visit colleagues. The ISDN call transfer services permit calls to a peripatetic executive to follow him around the world (CT, CFB, etc.). We need never again wonder who was that calling as we fumble with our key in the front door; the identity of the caller will be on the little screen on our ISDN telephone or terminal (CLIP, possibly CW). When we take our portable computer away with us and plug it in in a different country, the ISDN will recognize that we are at a new access of the worldwide network (TP).

That there are problems is undoubted. All the features dependent on the calling line identity are open to question because of legal objections to disclosure of this information, CLIP, CLIR, etc. All these services have to have in-built facilities to prevent identification information

Q.731 *Number identification supplementary services*
    Q.731.1   Direct dialling-in (DDI)
    Q.731.2   Multiple subscriber number (MSN)
    Q.731.3   Calling line identification presentation (CLIP)
    Q.731.4   Calling line identification restriction (CLIR)
    Q.731.5   Connected line indentification presentation (COLP)
    Q.731.6   Connected line identication restriction (COLR)
    Q.731.7   Malicious call identification (MCI)
    Q.731.8   Subaddressing (SUB)

Q.732 *Call offering supplementary services*
    Q.732.1   Call transfer (CT)
    Q.732.2   Call forwarding busy (CFB)
    Q.732.3   Call forwarding no reply (CFNR)
    Q.732.4   Call forwarding unconditional (CFU)
    Q.732.5   Call deflection (CD)
    Q.732.6   Line hunting (LH)

Q.733 *Call completion supplementary services*
    Q.733.1   Call waiting (CW)
    Q.733.2   Call hold (CH)
    Q.733.3   Completion of calls to busy subscribers (CCBS)
    Q.733.4   Terminal portability (TP)

Q.734 *Multiparty supplementary services*
    Q.734.1   Conference calling (CONF)
    Q.734.2   Three party service (3PTY)

Q.735 *Community of interest supplementary services*
    Q.735.1   Closed user group (CUG)
    Q.735.2   Private numbering plan (PNP)
    Q.735.3   Multi-level precedence and pre-emption (MLPP)

Q.736 *Charging supplementary services*
    Q.736.1   Charge card
    Q.736.2   Advice of charge (AOC)
    Q.736.3   Reverse charge (REV)

Q.737 *Additional information transfer supplementary services*
    Q.737.1   User-to-user signalling (UUS)

Table 46.1 ISDN supplementary services

being passed on from one public network to another. But some of the
facilities dependent on these, such as ordering and paying for goods by
telephone, depend on validation of the calling number plus, perhaps,
obtaining a credit rating from a central database.

Many of the problems and much additional complication within the
supplementary service protocols are associated with the requirement
that each use of the supplementary service can attract a charge. There
is, indeed, a strong argument for criticizing administrations for their
failure to offer ISDN access and ISDN supplementary services as a
single package with little or no additional charges for the use of the

individual services. Such a procedure would be much cheaper to implement and would make the ISDN a much more attractive proposition to the user. There is always a reluctance, as we saw in the UK with the failure of Prestel (see section 50c), for users to sign up for services whose charging arrangements are difficult to understand or to supervise and control.

A further problem, not unconnected with charging, is the problem of interactions between the different supplementary services. Completion of calls to busy subscribers (CCBS) may be affected in its operation by the presence of call forwarding supplementary service at the outgoing end. And if the two services complement each other will it be the calling user or the called user, or both, that will pay for the service? Similarly, under what circumstances is a member of a closed user group able to go outside the group when constructing a conference?

All these problems are being addressed; there is European standard ETS 300 195 on Supplementary Service Interactions. In Europe, the free availability of the ISDN, complete with supplementary services, is a major plank of the Open Network Practice Directives. All the Recommendations listed in Table 46.1 have their European counterparts as ETSI Standards. The conformance testing provisions for the ISDN are common to both the CCITT and to Europe. We have waited a long time but the ISDN is now available and its availability assured.

47 Mobile Radio Telecommunications Systems

The past decade has seen more innovation in the area of mobile communications than in any other area of telecommunications. Our quality of life has already changed as a result and there is undoubtedly a good deal further to travel.

It is no surprise then to find that the references to the various mobile communications systems are themselves very confusing. The object of this introductory section is to attempt to make sense of some of this confusion.

Figure 47.1 shows the common divisions into which the subject of mobile communications can be grouped. It is noticeable that we have omitted communications with ships and aircraft which have been dealt with already in section 33. They are undoubtedly a part of the mobile subject area and, as we shall see, they too will be relevant to our more land-based discussion.

All these services, apart from paging, have been introduced since 1975. This is the underlying reason for the confusing number of different

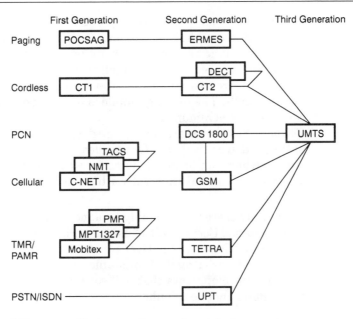

Fig. 47.1
Mobile radio system development history and future

CT1	First-generation cordless telephone systems
CT2	Second-generation cordless telephone systems
DCS 1800	Digital communication system at 1800 MHz
DECT	Digital European Cordless Telecommunications system
ERMES	European Radio Message System
FPLMTS	Fixed Public Land Mobile Telecommunications System, the CCITT name for UMTS.
GSM	Groupe Special Mobile, Global System for Mobile communications
Mobitex	Mobile data system (Sweden)
MPT1327	Ministry of Posts and Telecommunications (UK) Signalling standard
NMT	Nordic Mobile Telephone
PAMR	Public Access Mobile Radio
PCN	Personal Communications Network
PMR	Private Mobile Radio
POCSAG	Post Office Code Standardization Advisory Group
TACS	Total Access Communication System
TETRA	Trans European Trunked Radio system
TMR	Trunked Mobile Radio
UMTS	Universal Mobile Telecommunication Service
UPT	Universal Personal Telecommunications

initiatives. Complementary and competing services have been introduced to the market at substantially the same time. There has therefore been a confusion of overlapping and competitive developments.

A short description of each of the main services is provided here and a more extended treatment is included in the sections which follow.

Paging Systems permit the user, who carries a pocket receiver, to be alerted to the fact that he/she is wanted. In most cases it is necessary for the user to contact a central point in order to find out what is wanted. *Private Mobile Radio* provides two-way voice and, possibly, data communication between the user and a central point, or between users belonging to the same closed user group.

Cellular Systems provide continuous communication between mobile users and any other telecommunications terminal connected to the worldwide public network whether mobile or fixed. They are named after the technology of providing small transmitter/receivers serving cells so that the scarce resource of radio frequencies can be re-used in non-adjacent cells. The moving mobile is handed on to the cell best placed to continue the communication.

Cordless Systems started with the desire to have the telephone with you in the garden or the bath. The cordless' phone communicates with a base station which can be in the home or, with the use of Personal Communications Networks (PCN), with a base station in a public place.

Eventually there will probably be a merging of cordless systems and cellular systems. Certainly, the use of words like "total" and "universal" in the titles of fig. 47.1 is not yet representative of fact because of the variety of systems. Private Communications Networks (PCN) however do envisage developments that will satisfy the needs of cordless telephony and cellular mobile telephony.

47A Paging

Paging systems started off as very-low-power systems giving effective coverage over a limited area, say in the several buildings of a major hospital complex. Users carried, usually clipped to their belts, a receiver half the size of a paperback book. When they were needed on the phone or in the operating theatre a coded call went out from the central transmitter. This operated a buzzer in the receiver being called. The user then hurried to the nearest phone and called in to the operator to find out where he was needed.

As technologies advanced it became possible to provide smaller and smaller portable units, wider coverage areas and rather more information via the radio path. In later models a beep-tone buzzer sounds when you are needed, then a coded numerical group with agreed special meaning to you appears on a tiny strip-screen at the end of the receiver. More expensive models give a plain language message which, if it is too long to be shown completely on the strip, is read into the unit's memory and scrolled along the strip until you have read it all. For use in areas where a beep-tone warning would be unacceptable you can now get pagers which vibrate silently or flash a small calling lamp when your attention is needed.

In many countries almost nationwide coverage can now be provided by paging systems. In Britain paging services are available from seven wide-area operators including BT and Mercury. Most countries still use a system originally designed so long ago that BT was then still part of the British Post Office. The system is called POCSAG, from Post Office Code Standardization Advisory Group.

Pagers do not need to receive voice signals (although special pagers are in fact now available which do pass on voice messages) or to transmit any signals at all back to their base stations so the portable units can be relatively small and cheap and with low power consumption. They need only a narrow frequency band to pass their data signals. This means that many users can be given service even if radio frequencies are limited.

Paging has however proved so popular that congestion is now being experienced. A Pan-European service has been planned and is due to become operational shortly. This is a digital radio system called ERMES (for European Radio Messaging Service). When this is fully operational it will be possible with this one unit to page people all over Europe. There are already links between different national paging systems so it is already possible to page in the USA from Europe. This means it will soon be impossible to get out of range of your office if you are really urgently wanted.

47B Private Mobile Radio

If your business is such that you need radio contact with several mobiles—a fleet of lorries and vans, say—there are other forms of radio telephony which are much cheaper than having all your vehicles fitted with cellular radio phones. After all, your truck drivers do not need to be able to make unlimited nationwide or international telephone calls; they only need to be able to maintain contact with their operating bases and their base dispatchers need to be able to control their movements.

Until a few years ago the only possibility was a Private Mobile Radio Network of your own, with heavy capital and running costs and traffic handling constraints, especially if you needed coverage over a wide area of the country. Many organizations which operated transport fleets in fairly local areas, such as local councils, have often used such private radio systems. For countrywide use however, they can be prohibitively expensive. It has now become possible for many of the expensive features of this type of radio network to be shared with others on a Closed User Group basis. Different users share access to common frequencies, trunk communications links and the many radio stations needed to provide all the coverage that is required. This form of PMR is sometimes called trunked PMR. It should be noted that it does not necessarily provide access to the Public Switched Telephone Network, the PSTN; your mobiles are in touch only with your dispatcher.

Trunked services have been given a new lease of life in the UK by the allocation to their use of the "Band 3" frequency bands which were at one time used by the old 405-line TV system, now all replaced by colour TV at higher frequencies. In the UK there are several groups now providing PMR services, some are nationwide and some provide only regional coverage.

At present PMR is much the most economical way of keeping in touch with and controlling fleets of vehicles but in a few years time there may be competition from a satellite-based service designed originally for ships at sea. The Inmarsat organization has now developed a data-only receiver which does not need a complicated dish antenna. It looks just like a small car radio set and provides typed messages as its ouput (see section 33). Until these are widely available (probably in the mid-1990s) trunked PMR will remain first choice for fleet dispatchers on economic grounds.

47C Cellular Radio Systems

Cellular radio is a development that has surprised everyone by its popularity. A local development intended to overcome a local difficulty has provided the greatest success story since perhaps the invention of the telephone.

In the large stretches of thinly populated northern Scandinavia there have always been real problems with keeping in touch. Communications in these regions can be a matter of life and death. During the 1970s the Nordic countries co-operated in the development of a radio telephone system that would provide communications throughout Scandinavia without the need for an increased investment in the terrestrial network and without excessive use of the precious resource of scarce radio frequency spectrum.

The result was a cellular system based on the use of relatively low-powered base stations transmitting to and receiving from mobile user stations, usually provided as car telephones. The frequencies used in one cell could be used again in any but adjoining cells.

To everyone's surprise the system became most popular in urban areas for use by business people and there had to be almost immediate upgrading of the system to permit much smaller and far more numerous cells to provide adequate coverage in urban areas.

Very soon the system, or one like it, was in use in the UK, in other European countries and in the USA.

A cellular radio-phone is not much bigger than a paging receiver but far more complex, allowing ordinary local, trunk and international telephone calls to be originated as well as received. One of the difficulties with all radio systems is the hard fact that frequencies are limited. Good radio coverage of a wide geographic area at the very high frequencies used for these services can only be obtained by siting the main transmitters and receivers at as high a point as possible—on top of the tallest local building or on a nearby mountain top. This is fine so far as it goes but if you have available to you only enough frequencies to allow say 30 conversations at a time, whereas you need to be able to carry over a 100 calls at a time, then the mountain-top site with its wide area coverage is no longer quite so suitable. You need to be able to divide up your geographic area so that it may be served from more than one main station. Antennas may for example be made more directional, so that the one main site acts as if it were several different main stations, each covering say 90° of arc. By reducing the power of main stations, and in some circumstances placing their antennas only half-way up a building instead of at the top, it is possible to re-use frequencies in other areas not very far from where they are already in use. This means that the available frequencies, usually the resource which runs out first before all traffic requirements can be met, can look after far more simultaneous conversations than before.

This repeated re-use of frequencies is one of the main differences between cellular systems and their predecessors. The other is that mobiles on a cellular system are able to change their own operating frequencies as they move. As you move around a city the strength of your radio signal varies at the receiving station carrying your call. Whenever the signal strength gets too low for satisfactory operation, all neighbouring fixed stations are instructed automatically to listen out for your signals. If one of the neighbouring cell area stations receives you better than your original "home" station, control is passed over straightaway to this other station; your set changes its transmitting and receiving frequencies to ones used by the new master and the network control station switches all the external circuits over so that whoever you were talking to, whether a subscriber on an ordinary telephone system anywhere in the world or another mobile, is now connected to you via this new controlling station.

In Britain there are two competing nationwide cellular radio systems, Cellnet and Vodafone, each with its own network of linked radio main stations and switching centres. But by a somewhat complex series of regulations the radio-phones themselves are provided and installed in cars and "air time" is paid for via Service Providers and their local contractors, some of whom may be better than others. Some tariff structures appear to be more generous than others, but all calls via cellular are expensive. Some charge by the minute or part thereof for every call, some insist on contracts for several years service being signed before they provide any service at all. Calls to cellular phones from the PSTN are also much more expensive than ordinary PSTN calls.

One of the much publicized difficulties experienced by users of cellular services has been brought about by the great popularity of these services. There are now far more mobile phones on the road in some areas than the planned traffic-handling capacity of the networks can reasonably support. Cells are however continually being subdivided and new frequencies being made available in order to increase traffic-handling capabilities and remove congestion, but at peak times and in some areas it is still difficult to make a satisfactory and uninterrupted call from a mobile.

It is of course also not possible to use a present-day UK car-phone while you are driving anywhere in continental Europe and Continentals cannot use their car-phones here in the UK either.

A completely new digital cellular system has however now been designed which will provide Pan-European coverage. Users will be able to use their car-phones wherever they are in Europe and will eventually be billed for calls made. This new system, designed to replace all of Europe's existing analog cellular mobile systems, is called by the name of the Study Group which drew up its specifications, GSM (for Groupe Spécial Mobile). The developers and the makers of GSM equipment are now marketing GSM worldwide and have renamed it the Global System for Mobiles.

47D Cordless Telephones and Personal Communications Networks (PCN)

Many people buy cordless phones in high street shops to use as extensions so that they may answer their home telephone while they are lounging in the back-garden.

Difficulties are sometimes experienced with the present analog generation of cordless phones, especially with some of the cheaper imported models. For example the number of frequencies available for allocation is limited, so it could happen that a near neighbour might buy a phone set which uses the same pair of frequencies as your own units. If this happens, and your main set with its antenna is upstairs overlooking his garden as well as your own, his outgoing calls could sometimes go out on your line, at your expense. Or, even worse, there are people who might go to some trouble to make their outgoing calls over your cordless telephone. Users of such phones might well look out in the street to see who is parked nearby and talking on the telephone. The radio system for all these early cordless phones is ordinary analog and completely "in the clear". Anyone with a radio receiver covering the frequency bands concerned can listen-in, so nothing commercially or personally confidential should ever be said over one of these phones.

These problems led to the design of an advanced cordless telephone system called the CEPT's CT1 (Conference of European Posts and Telecommunications Administrations). The CT1 has now been followed by a new digital cordless phone called the CT2 for Second Generation Cordless Telephone. This is the system which has been used for the initial offerings of "Telepoint" services.

The CT2 designers, having licked most of the transmission and security problems, thought that the new CT2 was far too good to be used only by a few wealthy stockbrokers lazing by their swimming pools. The main stations in this new CT2 system are very small. They can be mounted on lamp-posts or in public places anywhere and your own pocket-sized CT2 will be able to pick up a telephone line through the nearest one of these tiny main stations and to identify itself by sending your Personal Identification Number (usually sent automatically by a chip in the set) so that you will be billed for the call in due course. You then key out the wanted number just as with an ordinary phone.

Telepoint services are not proving quite so popular as their developers expected. The UK, for example, had four systems initially planned but only one of them actually went into service and even that has not been very successful.

The major difficulty seems to be the contrast with cellular services. You cannot normally use one of these phones while you are on the move as the outgoing call is established through the particular main unit which has answered your "off-hook" calling signal. A further difficulty with this CT2 type service is that it is for outgoing calls only. The network, for economy reasons, is not designed to keep in its databases constantly updated records of the exact location of every possible user. If you already have a radio pager you can be advised via the paging system of the telephone number of the person who wishes to contact you. All you have to do then is to go to an area served by one of the many main CT2 Telepoint stations being installed in railway stations, airports and main squares and call him up yourself. CT2 phones are available which incorporate a pager.

Another bright idea then emerged—since it is now possible, through digital techniques, for very many low-power radio telephone links to operate simultaneously in a small area, why shouldn't this same type of procedure be used to provide a low-cost mass-market mobile, cheaper than the now "traditional" analog cellular mobile radio systems and using geuinely pocket-sized terminals appropriate to the higher frequencies used (2 GHz, against the 300 MHz of ordinary cellular)? This has now developed into PCN, Personal Communications Network, which uses radio procedures generally similar to those to be used by the 900 MHz GSM, the digital system designed to replace Europe's many different existing analog cellular radio systems.

Another system also being developed on a Pan-European basis is DECT, for Digital European Cordless Telephone (or Telecommunications). This started off as a modern replacement for ordinary analog cordless phones (i.e. the market which the CT2 was designed to penetrate) but the design has since been extended so that it can now be used for phones or data terminals in areas with very high concentrations of users, much higher concentrations than can be handled by GSM or by CT2. The idea is that radio links should be used not only by mobiles but should also be economic for fixed stations.

Each user could carry a tiny pocket phone on leaving the desk and be able to deal with calls even when visiting a colleague. This leads to what could perhaps be the biggest saving of all—why should it be necessary to plan and install complex underground cable networks all over our cities and to run cables in ducts all over our offices and homes? When a company moves its office base from one location to another or merely moves a few desks around to suit the latest reorganization scheme, the cabling schedules and cross-connection programmes needed are sometimes horribly complex and expensive. Why not carry all these telephone conversations and Personal Computer links on digital radio paths, needing no administratively expensive re-arrangements at all when the user moves from one desk to another? The cables still in their ducts, not then needed for ordinary telephony, could perhaps be used for a variety of broadband services, from high-speed data to video services. This is one of today's most exciting developments.

There is however some slight confusion in the marketplace at present because DECT will for some purposes be competing both with CT2s and with PCNs. Whatever does eventually happen however it seems clear that digital radio systems will be playing a major role in all our telecommunications systems before long, all of them developed from a wish to improve the humble cordless telephone. Remember here that the thousands of manholes and thousands of kilometres of ducts and cables now under city streets, all needed for our present-day telephone system, represent about half of the total capital investment of most national telephone networks, so their replacement by digital radio links, providing complete flexibility but without the expensive "burden of spare plant" necessary for buried systems, should eventually result in significant cost and price reductions.

48 Freephones: the "800" Series of Telephone Links

If a telephone call is free you are clearly more likely to make it than if it costs you a lot of money. That is why "800" calls (0—800 in the UK or 1—800 in America) are now very big business indeed. Like so many special telephone services this one started off in America. The idea is that wherever you are, your calls to an "800" number will be routed direct to the company concerned, and they will pay all the charges for the call, even if it uses expensive long-distance circuits.

"800" service calls cannot easily be provided unless the exchanges involved are computer-controlled; that is why callers in the UK sometimes have to dial "100" for the operator and ask for a "Freephone" call, instead of being able to dial the "800" number direct.

It will be remembered that until a few years ago each telephone exchange in Britain had its own name; in big cities you dialled the first three letters of this name (e.g. TEM for Temple Bar, HOL for Holborn), followed by the wanted customer's number.

When the expansion of the system made it impossible to compose pronounceable and meaningful exchange names, using the available letters, we in the UK went over to all-figure numbers. Most other countries did the same, at round about the same time, but not just because they were running out of numbers: international subscriber dialling was then beginning to become available and it is very relevant that different countries use different alphabets. How would a Russian, a Greek or an Egyptian dial-out letters which did not exist in their alphabets? And how would Japanese or Chinese or Koreans cope with WHI 1212? Even countries which use the same alphabets do not allocate letters to finger-holes or push-buttons in the same way; in the USA for example, O as a letter uses a different button from 0 as the number zero, whereas in the UK both letter O and number zero used the same button.

In America as in Britain we no longer use alphabetical letters in our published telephone numbers but in Britain the old letters have all vanished from modern telephone instruments whereas in the USA they are still marked on most dials and key-pads. This has made it possible in America for nation-wide "800" numbers to be spelled out in letters. Local florists for example can be reached anywhere in the USA by dialling 1-800-FLOWERS and many similar easy-to-remember codes are in common use. People can of course also use these letters as memory aids for ordinary numbers; for example with old UK-type dials, DARLING could well be easier to remember than 327 5464! Maybe we in the UK were a bit too quick off the mark in 'updating' our telephone instruments by removing the letters from our dials and key-pads?

49 Facsimile

Facsimile services are perhaps the oldest of the additional services available with the telephone. The original patent (granted to Muirhead) was intended for the transport of pictures between newspaper reporters and photographers and their editorial offices. Facsimile has been in use for this purpose since the early 1920s. The system scanned the picture and encoded its black through grey to white content, sent this code to line and the receiver reproduced the picture as a series of dots. The resulting copy was, and still is, of a remarkably high quality.

In the mid 1960s simpler, less-expensive facsimile machines began to be offered to the general public for use over dialled-up telephone lines. The use of fax has increased tremendously in recent years. Not only have speeds of transmission gone up but terminals have become more reliable. Standardization on an international basis by the CCITT has probably been one of the main reasons for this big jump in facsimile usage. The cost of preparing messages for transmission by telex has also influenced the growth of facsimile, which can handle rough manuscript drafts just as well as elegantly prepared typed copy. The use of fast digital transmission and the design of clever machines which rapidly ignore all the blank section on a page (and so give faster reproduction) are clearly helping in this popularization process.

Systems are now classified by the CCITT in 4 Groups:

Group 1   An analog system, about 6 minutes for an A4 page. This system is still sometimes encountered but most terminals have been replaced by newer ones.

Group 2   Another analog system, 3 minutes for an A4 page. There are still some of these around although they have not been sold for many years.

Group 3   Todays most popular, less than 1 minute for an A4 page. Group 3 machines can interwork with Group 2. They use a "handshaking" procedure to establish transmission speeds so if the distant end is Group 2, your Group 3 machine sends out more slowly.

Group 4   The latest model, designed to work over digital networks, not very common yet in the marketplace. When digital paths end-to-end are with us, as they will be when ISDNs are common, Group 4 machines are likely to sweep earlier models off our desks, especially in big offices. They only take about 4 seconds per A4 page. Telex and other data services may well find that fast fax overtakes them for many business purposes.

Fax machines of Groups 1 to 3 use a special form of thermal paper. More expensive versions of Group 3 machines are available using ordinary reprographic paper. Group 4 machines use ordinary paper and provide a copy indistinguishable from a good photocopy.

Documents on the special thermal paper used by most fax machines sometimes fade away within a few months, specially if they are stored in plastic files or exposed to excessive heat or sunlight. At present the best answer to this long-life problem is the use of plain paper machines or thermal transfer machines. These are however usually significantly more expensive than the more common types of fax machines. Some users, for whom long life of file records is essential, photocopy all their incoming faxes as soon as they are received.

Postal strikes and mail delivery delays, and the sheer cost of postal services, have led to the increased use of fax but most fax traffic at present (excluding Japan) is understood to be between head offices and branches of the same company. The marginal cost of sending a letter by fax can be less than the cost of sending it by post—and fax is faster and safer also.

In Japan, thanks largely to the fact that their ideographic written language does not lend itself to easily remembered telegraphic coding, fax is now widely used for rapid commercial correspondence between companies and this same trend is being encountered in many other Far East countries. Many firms in the East say they use fax more than telex or postal services.

50 Interactive Services

We have become accustomed to using the telephone to interact with a human user. If the person or device we called was busy or not available, that was the end of the matter. Nowadays we speak more often of "users" rather than "subscribers" indicating a shift in the perception of the user which can, as often as not, be a machine. In this environment it is convenient to discuss together the range of interactive services— services in which the user does more than make a call but is required also to interact, with information and requests, with the called user.

Telex, which we have already discussed in section 42, is an example of an interactive service; the other services of this nature which will be discussed in the sections that follow are:

Teletex   A service similar to telex but permitting high-speed both-way textual communications between terminals. This can be used for communicating wordprocessors or as a form of electronic mail.

Teletext   A service provided by the television channels where the user can request pages of information to be displayed from a menu. The UK versions of this are Ceefax and Oracle.

Viewdata   A service not dissimilar to Teletext where the user can request pages of information to be displayed on the television (or other) screen and can make simple requests such as purchases or the leaving of enquiries in a mailbox. The difference from Teletext is in the means of provision. The Teletext pages are continuously broadcast and the user must wait for the next broadcast of the wanted page. In Viewdata the request for the page and the page itself are transmitted over the telephone line.

Cards   The use of intelligent cards to validate orders to ensure the security of the communication.

EPOS, FPOS   Electronic funds transfer at the Point of Sale, Funds Transfer, etc. Perhaps the best description is Electronic Funds Transfer, EFT. The various arrangements to pay for and validate payment at the check-out desk.

Electronic Mail   All forms of data and text communications which imitate electronically the methods of document interchange that hitherto we have conducted with the help of the postman.

Voice Messaging   An extension of the telephone answering machine to provide a corporate facility including mailboxes for verbal messages to be left for any member of the organization.

EDI   Electronic Data Interchange, a term used to describe all those functions which will enable us to conduct our businesses electronically, ordering, invoicing and paying for goods and services in an immediate, paperless manner.

50A Teletex

We have here a difficulty in recognizing similar names for very different services. Teletext with a "t" on the end is the broadcast data service transmitted by many TV stations without interfering with their ordinary TV transmissions. This is dealt with in the next section.

Teletex, with no final "t", arose out of the provision of packet switched data services (see section 35), and is a super form of telex giving upper case and lower case characters and many signs and symbols which telex cannot manage. It is very fast, only about 8 seconds for a normal A4 size letter compared with minutes for ordinary telex, and it will take less than 1 second per page when networks have all been converted to digital and ISDN is generally in service.

Germany has had a teletex service for several years; for various reasons it was not provided in the UK until 1985. There are still many countries where teletex is not available and it would not seem unfair to say that the whole future of teletex is at present somewhat cloudy. The terminal looks rather like a modern telex machine, with VDU screen to show the message being prepared, but the electrical and mechanical sides are quite different. A teletex machine in some ways acts like a computer terminal: when you connect it to the distant-end machine the message is very rapidly transmitted from the memory store at one end to the memory store at the other, then brought up at your convenience either to the VDU screen or printed out on paper.

Teletex gets its extra speed by using a much greater bandwidth, usually 2400 bit/s compared with only 100 bit/s or so for a telex machine. This 2400 bit/s can however be carried by ordinary local telephone, no special lines are needed. Teletex services are normally carried by national packet-switched networks between towns.

Terminals on the teletex network can interwork with machines on the telex network. The difference in speed is taken care of by the teletex system itself which holds messages in reservoir stores until they can be delivered to their slower-speed cousins. Teletex quality is also so good that it has encountered one unexpected difficulty. No-one expects to see a signature at the end of a telex message to confirm its authenticity, but teletex messages look just like high-quality typed letters so some recipients feel that the absence of a signature at the bottom in some way makes it look as if the message may be an unapproved draft, not a genuine message.

Perhaps the failure of Teletex to achieve wide popularity is partly due to the fact that the same service is effectively available between any two PCs equipped with a dial-up modem giving them access over the public telephone network or, better still, over the packet switched network. With the coming of ISDN access the communicating PCs will not even require the modem and will be able to work, if necessary, at even higher speeds.

50B Teletext

If W. S. Gilbert were compiling his list today of all those people "who never would be missed", it is to be hoped he would include the light-fingered user of the TV remote control who looks up Teletext to see what is on the other channels just when we are engrossed in the show. Just so does technology present us with fresh annoyances alongside new developments.

All the pages of information in Teletext services are transmitted by TV broadcasting stations together with their normal signals. The teletext pages are sent out using the 20 lines of the "blanking interval" which are not used for the ordinary TV picture itself. Each blanking interval is capable of carrying 14.2 kbits of data using teletext code standards. All the pages of information are transmitted on a cyclically repetitive basis so there can be some delay between choosing a particular page number and the display of that frame on your screen. To minimize this delay most modern TV sets include memory units which store the most commonly called-for frames so that these pages will normally be displayed immediately their number is keyed in to the set.

In addition to news and sports coverage teletext services provide subtitles for most popular TV programmes so that the hard-of-hearing can read what the characters are saying.

From a business point of view, teletext services are of interest for their advertisements—when TV programmes themselves are not particularly absorbing it is common for viewers to switch over to teletext, consult the index and key in the numbers for the various advertisements to see if there are any bargains on offer. Some agencies indeed claim a greater response per advertising pound from Teletext adverts than from any other form of advertisement.

50C Viewdata

Viewdata, also called Videotex, was a British first but like many such firsts has been less successful in its country of origin than elsewhere. Prestel, the British service, was introduced in 1979.

Prestel service is provided in Britain by a network of computers organized to store frames of information which are transmitted (over the telephone line) at 1200 bit/s when requested by the customer. A reverse channel at 75 bit/s is provided to allow the customer to select the frame required or to provide answers to questions set in the information frame itself. This bit rate is sufficient for manual keyboard operation. Fig. 50C.1 shows a typical terminal.

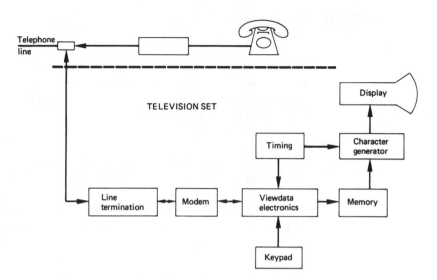

Fig. 50C.1
A typical viewdata terminal

The information on the database is provided by a large number of specialist companies. Some pages of information have to be paid for, some are free. Some companies put information on the database on a restricted access basis, e.g. only Prestel customers from their own organization are permitted to call up these particular frames.

One major difficulty at present is that the viewdata systems in use in different countries use different procedures, codes and signals so they cannot all interwork with each other: customers using British Prestel cannot, for examples, access the Canadian Telidon system. The Prestel system was the first of its type to be developed and uses what is called an alpha-mosaic method of constructing letters from small squares of colour; that is why they seem a bit odd to those who are used to normal TV definition pictures. Telidon, much later in the field, uses an alpha-geometric method of constructing letters, with lines and curves as well as squares of colour, so it is able to give an easier-to-read page of text and better diagrams than Prestel. Some countries are now experimenting with hybrid viewdata/teletext systems by which the interactive or "command" signals from the customer to the computer are sent in on a telephone line but the called-for page of information, either test or picture, is transmitted by a radio signal, coded, stored and displayed only by the customer asking for this particular page.

One of the factors which may have caused the failure to achieve a sufficient market for Prestel may well be the method of payment. One has to pay for the rental of the telephone line and the call charges, one has to buy or rent the Prestel terminal, one pays the Information Provider for viewing particular screens of information, and users wishing to contribute information must pay to have their screens displayed.

France Telecom adopted a completely different approach. They argued that the saving in not producing free directories could justify the distribution of a cheap viewdata terminal to access directory information, and the system would come into profit by means of the charges made to other bodies placing information on to the French system called Minitel.

Perhaps another factor, which dogs all existing systems, is that the system is only as good as the information it provides. If expected information is not there or obviously out of date, users lose confidence in the service. Personal experience has shown that this is a problem even in the best of the existing services, that of France Telecom.

In the UK the most successful users of Prestel have been travel agents and most present-day agencies are equipped with Prestel terminals.

One development of Viewdata, to carry pictures of acceptable quality, much wished for by estate agents, has been overtaken by the ISDN. In France the ISDN, Numeris, already has the equivalent to the Minitel service operating and this includes the transmission of high-quality pictures.

The provision of Viewdata services becomes, technically, a very different matter when allied with cable television (see section 51). The communication with the viewdata databases could be much more simply provided down the cable along with the television signal. In fact we then would envisage a service like the BISDN provided over the cable TV network as suggested in section 28. This is an illustration of how all the manifold services which have been made available to us piecemeal over the past 20 years become a part of one universal service over a universal access.

Just as Prestel failed in Britain and succeeded (to a point) in France because of different approaches to charging, much of the delay in achieving the universal access of the ISDN and then BISDN are caused by an unimaginative and grasping attitude to the problems of charging for so many diverse services.

50D Credit Cards and Smart Cards

Metal coins and paper currency may be a bit of a nuisance (although a wallet full of crisp new high-value notes would no doubt be very acceptable to most of us) but in the more developed countries of the world there is a well-established move away from the use of hard cash. Major payments have of course for many years been carried out using cheques or bank drafts, but there is now also a steady increase in the use of "plastic money", credit cards, for relatively small payments.

A high proportion of hotel, airline and restaurant bills in internationally popular tourist areas is now paid by credit cards and the use of such cards for ordinary retail purchases has now become generally acceptable in many countries: so much so that in some areas of the USA a man who pays cash for goods is somehow rather looked down upon, he is thought to be someone who must have been turned down by credit card organizations as not being credit-worthy, not a good-risk!

When a purchase is made using an ordinary credit card (American Express, Diners, Visa, Master Card, etc.), the sales assistant sometimes vanishes for a short while; she has had to make a telephone call to the office of the credit card company to make sure that the particular charge concerned may be accepted. Most ordinary credit cards do however usually have a backing strip of magnetic material. When the card is placed in a special reader, the card number can be read out, automatically. In many countries, special telephone instruments are now available. The credit card is wiped along a slot in the back of the instrument; this automatically sets up a call to the credit card company concerned and obtains the necessary authorization without the shop assistant having to carry on a whispered—and possibly embarrassing—conversation.

When Integrated Services Digital Networks (ISDNs) have been brought into service, it is possible that services such as authorizations of credit card sales will be carried out on a packet switched basis. The small amount of data to be transmitted and received does not need the establishment of a full speech path, the 56 kbit/s or 64 kbit/s of a normal telephone call set-up; data at 8 kbit/s is perfectly adequate for these "Point of Sale" (POS) transactions. When ISDNs have become widely available, POS instruments able to read credit cards, send coded information straight to the card company computer, and receive its reply, all automatically, should overcome many of the difficulties now experienced with credit cards both by consumers and by retailers. The huge lists of lost, stolen and invalid cards, almost as big as some telephone directories, will for example no longer have to be ploughed through before a $10 purchase can be agreed.

But all these transactions still use paper as their final authority, a signature is still needed on the credit card payment slip.

Some telecommunications administrations, faced with huge losses by the vandalism of public payphones (coin box telephones), have introduced semi-intelligent cards which can be purchased for cash and which can be inserted into a slot in a payphone to permit calls of up to a predetermined total cost to be established. The balance still available after a call has been completed can be read off the card by the payphone into which the card is next inserted, and so on until the whole amount has been used up. The card is then thrown away.

It is only a small step from these payphone cards to what are now being called Smart Cards. These look like ordinary credit cards but they have a built-in electronic memory and the logic circuits of a microprocessor controlling read and write access to this memory.

The true Smart Card demands a universally acceptable standard which is now being finalized (in Europe by the European Telecommunications Standards Institute (ETSI)). Smart Card payment is already possible in limited areas. In France, for example, payment of motorway tolls by monthly account is possible using an intelligent device mounted on the vehicle windscreen.

In France also, the humble telephone card is a form of simple Smart Card with the debiting algorithm performed on an integrated circuit clearly visible on the card.

The card issued to users of Sky Television in the UK is also a form of

Smart Card having the key to the descrambler, required for the receiver to make sense of the scrambled program information, contained in the card.

Trials are under way of several different types of Smart Card; there is as yet no universally acceptable standard. For use in ordinary retail transactions, the Smart Card acts like an Electronic Cheque Book. The user has to remember (and key in) a Personal Identification Number (or PIN). The card's memory is programmed to know the total value of all purchases that may be made during each month. It authorizes and records all transactions up to this limit, and by placing the card in a special electronic reader, available at Point of Sale, the holder can see a display of current outstanding credit. This preset "revolving credit" limit is restored monthly. When a purchase is made, the POS terminal in the shop records all relevant information. Then when the day's work is over or at any other convenient time, all the data is transferred automatically to the bank for crediting the shop's account and debiting the customer's account. It can be seen that this will not only cut administration costs but will also make Smart Card sales much more acceptable to small money-hungry shopkeepers than credit card sales are now. In some countries it can take three weeks or even more before payment is actually received by the shopkeeper from the credit card company, so discounts which are freely available for cash purchasers are sometimes not given to credit card users. This naturally holds credit card sales down.

By the end of this 20th century many of the big shops in most developed countries could well therefore be operating on a completely non-cash basis. There would be no need for citizens of rich developed countries to carry pockets full of heavy money. A reserve supply of a few small coins (to permit casual purchases of candy bars and magazines) is all that will be needed to back up their Smart Cards. For many people in third world countries of course (and for many in rich countries also), even a pocket half full of money will, regrettably, still be beyond their wildest dreams but this will not be the fault of the telecommunications engineer!

50E Electronic Funds Transfer at the Point of Sale (EFTPOS)

The subject of cards brings us to the use of those cards and other means to effect cashless transactions at the check-out desk and over the telephone.

This has been possible in the United States since the late 1970s but has only become familiar to us in Europe in the past five years. The early introduction in the US was due partly to the much greater use of credit cards there and partly because of the generally much cheaper rates for the long-distance telephone calls required to validate the transaction.

In Europe, the introduction of the credit card and the automatic bank teller machine first made us accustomed to carrying our card about with us and this has now been extended to card reading and validating terminals associated with many check-out tills.

The technology required for all this is just a duplex rather slow speed data communication of very short duration. This is ideally suited to the packet switched network or to use of the D-channel of the ISDN basic access.

Advances in technology can be used to improve on the present, rather rudimentary security provisions. Most cards presently rely on the user keying in a remembered PIN number. Interactive, intelligent cards, allied perhaps with electronic transmission of a facsimile of the signature, can improve security. More complicated measures, such as the updating of a code held on the card on every validated application, have the disadvantage that the card becomes useless the first time the user miskeys the PIN number.

The fact that each transaction can now be immediately verified and funds transferred at once has changed the business prospects for the original credit card. Many users prefer the immediate transaction, without credit and therefore interest payments, and very possibly with a discount from the store for immediate payment. Without EFT this was not possible; the delay in payment was integral with the security provisions for a card where just the signature was validated.

50F Electronic Mail

Some of the earliest computer networks and all modern local area and wide area networks include a facility for the users to send each other messages. Each user is given a mailbox and every time the user "logs-on" to the network an indication is given of the messages waiting in the mailbox. Used wisely, this facility can revolutionize office procedures. Instructions and guidance can be provided to the people concerned and responses obtained with very little delay. Just as, however, we are used to seeing towering in-baskets awaiting attention in our colleague's office (never our own) the mailbox can become a cemetery for all our hopes of efficiency if it is overloaded with unnecessary memos or if the users fail to read it and deal with it regularly.

The same idea of electronic mail has been extended to the worldwide public network and operates in most of the major business private networks.

The cost of collecting, sorting and delivering ordinary mail continues to rise; these processes are necessarily somewhat labour intensive in most countries. Such mail does not have to be delivered within minutes of its dispatch; we are usually more than happy if letters are delivered during the day following their being posted. It has been calculated in the USA that only about 30% of their first-class mail is personal in nature; about 70% represents business correspondence such as accounts, orders, receipts, invoices, payments, etc. Almost all these are prepared by a computer in the company of origin and are dealt with by a computer in the company to which they have been delivered.

If the telephone network is to be used for the transmission of material which does not necessarily have to be dealt with on a real-time basis, there will have to be an indication given of the nature of the office traffic to ensure that telephone calls themselves are put straight through. These other lower-priority tasks can then be dealt with in bursts of information whenever the circuits would otherwise be idle. The type of signalling facility needed to control such services is already available with CCITT's common channel signalling system No. 7. This can be used to control the setting-up and supervision of all types of calls, both voice and non-voice.

It seems likely that for a good many years yet, very few households will find it economic to equip themselves with special apparatus to receive electronic mail direct. Such mail is likely to be printed out on high-speed printers in the local Post Office and delivered by messenger. In the business world, however, electronic mail (E-mail) has already taken off and is growing rapidly. Much of this development followed the popularization of Personal Computers, PCs. With a simple modem to enable a PC to send data over an ordinary dial-up telephone circuit, it is now common for messages to be sent direct from one PC to another, in another city. With more traffic the use of packet-switched networks becomes more attractive, with messages handled on a completely digital basis all the way from one end to the other.

It is a salutary thought that a human voice signal carried on a PCM system is represented by 64 000 bits/s. If a telephone call lasts 3 minutes, almost 12 million information bits will have been sent in each direction during the call. A typical bill being sent by electronic mail is, for comparison, likely to need only 3 or 4 thousand bits, and these could be transmitted whenever the channel is idle. It will be very easy to use national telephone networks to deliver such mail electronically, far more speedily than it can at present be delivered.

For truly universal electronic mail systems, as for any other form of communication, all the users must be using compatible protocols at all seven layers of the OSI model (see section 34). This is now possible for electronic mail through the use of the CCITT X.400 series of recommendations describing the message handling systems which define the Layer 7, the application layer for universal E-mail. Increasingly, organizations installing local and wide area network systems are specifying X.400 compatibility.

50G Voice Messaging

Most of the famous messages in our history books are one-way signals to which no effective answer was possible—the simple message "NUTS" issued by a US General when the Germans invited him to surrender at Bastogne in 1944 is a classic example of this.

The majority of telephone calls (apart from purely social gossip calls) are originated because people either want to tell someone something, or to ask someone to do something. There may of course be lots of interactive chit-chat mixed up with basic requests, questions and answers but the main reason for each call is usually fairly specific.

Telephone people all over the world are perhaps a bit like confidence tricksters: we have persuaded others that they really do need both-way conversation, when all most people basically want (in the business world at any rate) is to be able to deliver a message which will be acted upon. If the distant-end person is difficult to contact, it could be that, by the time you get through, the person has just gone out to a meeting. The return of your call finds that you are with the Chairman and cannot be disturbed. Your return call in its turn gets through just after your friend has left the office and is on the way to Miami for the weekend—all good business for the telephone company but not really very productive for you!

The ordinary telephone answering machine was an early way of dealing with this difficulty; remember that the tape recorder is a relatively new invention, it was not until the 1960s and 70s that these became widely available. An answering machine is connected to your personal telephone so that, if you do not pick it up after the third ring or so, it starts up and informs the caller, usually in your own voice, that you are away until tomorrow, please call back later. The next refinement was to equip the answering machine with the ability to record: "please say who you are and leave your message as soon as you hear the 'beep' tone". When you get back to your office, you play the whole tape back (or your secretary types out a list of callers' names and messages) so that appropriate action can be taken. With more sophisticated machines you can call in from outside and command your machine to play back the received messages to you, over the phone.

Strangely enough, overall public reaction to answering machines is not usually particularly enthusiastic. People who have them say they are useful but a great many callers somehow seem shy about talking to a recording machine; they just mumble "sorry", and hang up.

Answering machines become especially useful if there are time zone differences: a businessman in Singapore has a sudden thought to pass to a colleague in Chicago, but it is 17.00 in Singapore and only 03.00 in Chicago so an ordinary telephone call would not be very popular, unless of course big money was involved. Much better to phone this message through to a machine in Chicago which will pass it on at a civilized hour later in the morning.

A user of a voice messaging system (VMS) is pre-geared to talk to a machine. The psychological hurdle (associated with ordinary answering machines) of expecting a human reply but getting a click and a non-interruptible robot does not have to be jumped. VMS users can certainly expect lower long-distance call bills than those who insist on person-to-person contact. That basically is what todays voice messaging systems are all about. VMS rationalizes the concept of the telephone answering machine and makes these facilities available either to all company officers (for in-house messaging systems) or to anyone who joins a particular commercial Voice Messaging Service. You dial the number and speak your message. The system tells the addressee that there is a message waiting in the "mailbox", and by keying in a command digit, the message is played back and then erased.

One difficulty facing the marketing executive of a VMS supplier is the fact that the VMS primarily saves money for the caller who gets a message delivered regardless of whether the person called is there or not and who doesn't spend time and call charges fruitlessly listening to ringing tone. It requires an exercise in imagination for the company to realize that saving its callers money and getting its business performed regardless of the whereabouts of its staff is good economics.

The voice messaging mailbox is exactly similar to the electronic mailbox discussed in the last section. In fact, there is some difficulty in seeing when one could actually get any work done with all the pressure of keeping abreast of the mailboxes. It is probably true to assume that an organization with a good electronic mail system does not need VMS as well and vice versa.

Section 44 dealt with direct dialling in to the private network. It is certain that an organization using direct dialling in ought also to use a VMS system as a back-up. Otherwise, unless the staff are very careful to use their fall-back call transfer options there is a very real danger that the organization will gain a reputation for never answering calls.

One interesting adaptation of the VMS has been the introduction of home banking services. The bank customer has to be equipped with tone dialling in order to input the PIN number and the user's requests for service are recorded on the VMS for later implementation by the bank staff.

50H Electronic Data Interchange (EDI)

1 EDI, Computer to Computer

EDI was a subject much promoted by the UK Department of Trade and Industry in the middle 1980s. The force of their propaganda was that the use of EDI could result in paperless offices. All the normal business transactions of issuing an order, preparing the invoice and delivery documents, and paying the bill could be conducted by computers and word-processors interchanging documents over the public network, either the PSTN or the packet switched network.

Now that many national administrations are changing their transmission networks over from analog (FDM) to digital (TDM) working, it is comparatively straightforward to arrange for one computer to work direct to another. Why go through the stage of printing a bill, to be sent physically through the mail, when the information could easily have been passed from one company to the other, without being manually handled? In some organizations it may not in fact be necessary for a paper copy of a bill to be prepared at all; any necessary checking and passing for payment could be done using a visual display unit and a remote terminal on the company's computer.

For EDI to be successful, all the parties to the arrangement must be using the same standards to send and interpret the data. All the parties may well be using different computer systems and different data processing systems. How then can they communicate computer to computer without any direct human intervention? The parties must all subscribe to the same network, agree a message format, and load appropriate software to provide translation services, EDI services and network access services. These might operate as follows:

Company A's computer system produces a message (order for 200 widgets) and passes it to the translation service software. This translates the message into a common agreed structure and sends it to the EDI service software. EDI service creates the commands necessary to send the message, track it and ensure that it reaches its destination. Network access software handles the actual transportation of the message across the network.

In the early days EDI was carried out by groups of companies trading within a common industrial area. These groupings of companies created their own message standards and often supported their own private networks to convey the EDI messages. This led to the creation of a number of distinct national and sectional standards defining the structure of the EDI communications.

The switched data network of the Society for Worldwide Interbank Financial Telecommunication, SWIFT, for example, enables banks to send payment instructions to each other in a quick, error-free way in formats that each bank understands.

In one way this could be considered to be indicative of the type of service which EDI could provide for many other industries—SWIFT aims at providing a secure, standardized, auditable, controllable and rapid method for member banks to effect financial transactions. Because messages often deal with huge sums of money, security and control receive special attention, more than would normally be appropriate for an electronic mail system. For example, if an urgent message has not been delivered to its addressee anywhere in the world within 20 minutes, the sender receives an "overdue" warning advice explaining the reason for non-delivery.

The trend today is for these groups to come together and use internationally recognized message standards. The major standards setting body in this area is UN/EDIFACT (United Nations/EDI for Administration, Commerce and Transport). EDIFACT comprises a set of internationally agreed standards for electronic interchange of structured data, particularly that related to trade in goods and services between independent computerized systems. The other standards groupings have committed to harmonize with EDIFACT. At the present time however most North American EDI interchange uses the standards developed in the ANSI (American National Standards Institute) X.12 protocol.

In all applications of EDI there do have to be fairly careful controls built in to the systems to ensure that only real transactions are initiated and completed. There is much evident opportunity for the criminal or careless introduction of false transactions.

2 Public Access Databases

There is another kind of electronic data interchange which we have not already discussed. This is access to publicly available databases. The information technology equivalent of the dictionary, the trade directory, the library. Here again we have been close to the subject in discussing the databases available with Viewdata.

Many organizations collect a tremendous amount of information and store it in central memory units for the use of their own staff—a corporation's database of this type can usually be accessed from remote terminals (by those who have the necessary password) and the information in it made readily available to all authorised users. No-one knows who had the next idea first, it probably occurred to many people all over the world at more or less the same time: if all this information proved useful to their own employees might it not be useful also to others, to outside people who would be willing to pay for the information provided? Any part of the database which included "company confidential" information could of course readily be barred to outsiders. This same sort of service was then built-up on a completely public access basis by several libraries and by many other organizations which made the collection and selling of information their main business, and a very useful (and profitable) one it has become, specially in the USA where annual sales by databases now exceed 2 billion US dollars.

Anyone can establish an ordinary telephony call to one of these databases. There are more than 2000 of them, all listed in readily available special directories. When you dial a database number you hear ordinary ringing tone but when the database answers there is no human voice there, the answer is usually a high-pitched tone. You then have to tell the distant end who you are by typing in (or automatically transmitting if you have an intelligent terminal) an access code, your Personal Identification Number (PIN code) and your password. All these are necessary because no computer is likely to tell you all its secrets unless it knows who is going to pay the bill. You usually then get a menu on your VDU screen so that you may select the particular items of information you require. If, for example, you want to know if anyone, anywhere, has already patented a device which you have yourself just invented, you can get through to a database which specializes in Intellectual Property information, provide a few keywords, and it will respond by giving you brief descriptions of all valid patents mentioning these keywords. You can then go round to a local library or to the Patent Office Library and examine these patents for yourself in detail to see if your invention is really something new or if you have been beaten to it.

The prices charged by ordinary commercial databases vary widely. In the USA there is usually an initial registration fee of about 100 dollars. Most then charge by the minute with rates varying from a few cents up to several US dollars per minute, sometimes with different charges depending on the type of modem you are using and the speed at which you are working, e.g. a service at 1200 bit/s may cost you twice as much as one at 300 bit/s. Some charge also by the number of items you retrieve and have special (very high indeed) rates for information which they consider to be of great commercial value.

Part H
Television Services

51 Cable Television

Cable TV started off many years ago as CATV, Common Antenna or Community Area TV. It was a way of providing reliable many-channel TV services to people living in areas where the reception of ordinary broadcast TV was poor. A well-equipped receiving station was built on a nearby hill-top and cable laid to all the houses in the area, permitting them to choose from the many TV programmes which the hill-top site could receive. The same principles were then found to be relevant also in many highly developed down-town areas, in the USA in particular where steel-framed skyscrapers made ordinary TV reception very poor. These became prime markets for cable TV which therefore became popular in both cities and in rural areas.

In continental Europe also there are very many multi-channel cable TV systems. Several countries do, for example, make BBC and ITV programmes available for their viewers in addition to outputs from other nations' organizations. Britain is something of an odd-man out here; possibly BBC and ITV coverage and programmes have been generally thought to be all that people really need.

The advent of satellite TV is giving a new lease of life to Cable TV systems in the UK. The availability of Sky channels received via the Astra satellite and the possibility of picking up and relaying by cable programmes originated all over the world is an exciting prospect. The entertainment industry is now very much in the driving seat and satellite and cable systems are here to stay.

This pressure from the entertainment side of the industry is tending to divert attention away from return path usage. In the 1970s it became possible for most cable TV stations to provide a reverse channel. This was used, if at all, for polling audience reactions to programmes, to get votes on viewer preferences. The full potential of the reverse channel was rarely utilized.

With the advent of optic-fibre-based cable systems it was realized that this reverse channel was a very valuable property. It was a complete waste for it to be used solely for a few Yes or No signals in reply to marketing questionnaires. It was able to transmit speech, music or even TV pictures, so why not use this two-way ability inherent in these new cable TV systems to connect up ordinary two-way devices like telephones, or other more exotic devices like Viewphones, to be available for use whether or not the viewer was watching any of the entertainment channels on cable TV.

Cable TV with full two-way facilities is however a bit of a slow starter at present. Until about 1983 it looked like being the hottest development around. Several continental countries already had multi-channel (but one way) cable TV systems and were talking of converting these to two-way operation, and there were stories in the UK's popular press about the proposed "re-cabling of Britain". But then the financial climate changed and in effect these high-technology cable schemes were seen as relatively poor investments, too risky to be undertaken at that time.

Encouraged by the prospects of providing the local telecommunications network integrated with the CATV network, British Telecom performed quite a lot of pioneering developments which culminated in a major field trial of ISDN provided via a CATV network in Bishops Stortford. Before the trial was in operation however the BT licence had been amended by OFTEL to ban them from becoming involved in CATV/telecommunications developments. This is because BT, the monopoly provider of the local telephone network, was thought to be a danger to competing network providers if it were allowed to compete in CATV networks. As no other provider has attempted to challenge the BT local network monopoly (because it would not be economic to do so) and BT has therefore to provide the network for its rivals, this decision seems a little unjust. Its result has been to slow drastically the development of a common CATV and telecommunications network.

It has to be said that BT has not taken the alternative course that is still open to it, that of negotiating with a CATV provider to carry the BT local network over its cables. By contrast, BT's main rival, Mercury, has offered its network services via CATV and is involved in this way with several CATV networks.

In many countries with no competing telephone services the real sticking point in the use of cable TV networks to serve telephones also is the fact that existing telephone networks represent such a huge investment, and about half of the total costs represent the underground distribution networks. So no-one wants to devalue these (mostly Government owned) assets to zero, about all they would be worth if cable TV's network took over the distribution of our telephone systems instead of this being done by the existing network operator. Really modern interactive cable networks, providing us with all the entertainment we need as well as with all the telecommunications services we need, are therefore likely to become a reality first in those countries committed to BISDN and fibre re-cabling of the local network.

Another of the factors which affects the economics of using cable TV systems for ordinary telephone services is the fact that cable TV is at present primarily geared up to provide entertainment. This means that cable TV distribution networks must, for good economic reasons, concentrate on serving the residential areas of our towns, not the central business districts, which are likely to want plenty of telephones but not many entertainment channels.

52 High Definition Television (HDTV)

In planning the fourth edition it was decided to exclude the material in earlier editions not directly connected with two-way user-to-user communications. This section, on high definition television, has been spared because it will be delivered, and its pictures will be used, as part of the advanced communications arriving with ISDN and BISDN over fibre cables into our homes and offices.

Television picture standards have not been changed for many years. In the USA the National Television System Committee (NTSC) produced the first colour TV standards in 1953; these are still in use in North and Latin America and in Japan. Europe was later than America with colour TV; in the 1960s the Germans developed their Phase Alternate Line (PAL) system as an improvement on the NTSC system, and at more or less the same time the French developed their Sequential Couleur a Mémoire (SECAM) system. The two systems are in wide use in Europe, Africa and Asia. These two European-developed colour systems offer considerably better resolution and colour quality than the original NTSC system; 10 years made a tremendous difference.

Twenty more years have now passed. If a colour TV system were to be designed and manufactured today without reference to the standards of the 1950s and 1960s, it would undoubtedly be a very different system from NTSC, PAL and SECAM, and would provide picture quality as good as that of the best colour movies.

The use of HDTV practices is in fact likely to become common in the moving picture industry before HDTV really hits the broadcasting world. The reasons for this are that electronic editing at computerized consoles is much cheaper and quicker than having to develop film and cut it. Titles and special effects can be added electronically far more cheaply than by using traditional procedures. The use of HDTV technology in the movie industry is, however, in effect a closed-circuit use within the industry; it does not affect TV broadcasting stations, bandwidth economy or home TV receivers.

The NTSC system needs a bandwidth of 6 MHz; PAL and SECAM both need 8 MHz bands. For comparison an interim HDTV standard has been proposed by the European Eureka group, based on 1250 lines and an aspect ratio of 16:9. Such a system requires wider signal bandwidth of nearly 30 MHz and uses a radio bandwidth when transmitted by FM radio of some 100 MHz.

The use of HDTV display techniques by national ISDN systems is also under close consideration. When optic-fibre local distribution networks become available, the higher bandwidth needed for HDTV will present no major difficulties: it will then become possible for top-grade colour reproduction on big screens to be available in any home.

Since TV picture quality in NTSC countries is worse than that in PAL/SECAM countries, and since the electronic industries in the USA and Japan have traditions of being able to take extremely rapid action, there could well be more pressure to introduce HDTV in Japan and the USA than in Europe. All that can safely be said at this time is that those who are first into large-scale production of HDTV will doubtless set standards which the rest of the world, coming along later, will either grudgingly be forced to follow or will ignore, hoping that the more time that elapses the greater the chance that it will be possible to agree on a single standard to be used by all countries of the world.

Slightly higher picture quality was provided when BSB (British Satellite Broadcasting) began using the MAC (Multiplexed Analogue Component) system which was designed in the UK and accepted by Europe as the colour system to be used in future on all satellite TV systems. However, the hard facts of the marketplace stepped in and forced BSB out of business, leaving Sky as the victor despite the 30-year old picture quality, proving perhaps that viewers do not regard picture quality as of high priority.

53 Direct Broadcasting by Satellite

Early telecommunications satellites had very little power: in order to pick up their signals (remember that these have to travel nearly 36 000 kilometres to reach the earth) very-large-diameter high-gain dish antennas were needed. Ground-stations working to Intelsat satellites still need dishes at least 11 metres in diameter. (These stations do, of course, send signals up to the satellite as well as receive the signals down from the satellite.)

With improvements in technology it is now possible for satellites not only to put more power into their transmitted signals but also to feed this power into specially designed antennas on the satellites themselves which focus all the power on to a comparatively small part of the earth's surface. (Early satellites had to accept that their transmissions were directed towards the whole face of the earth.) This ability to focus radio powers on to a small ground area means that a receiver on the ground no longer requires a huge dish antenna. A very small unit only a few centimetres across is perfectly adequate to give a good signal/noise ratio and an acceptable TV picture. These small antennas are, of course, for receive-only stations; it is not possible at present to transmit television signals back up to a satellite using tiny dishes. It is, of course, possible to transmit data.

The use of this later generation of satellites (with higher power, greater bandwidth and special directional antennas) has therefore now made it possible for satellites to be used to broadcast programmes direct to users. Every home may well soon have its own small dish looking up at a geostationary satellite poised over the equator, beaming TV programmes down to continents and countries.

European countries are currently served by some 35 satellites. Earlier editions of this book illustrated the satellite coverage but the reader is better advised to consult a more up-to-date reference as the situation changes so rapidly. A current reference is contained in *Cable and Satellite Europe* for March 1992. While national policies on programme and advertising content differ, it is proving difficult to enforce these policies on the satellite broadcaster. While the possibility of enforcement on cable television relays is greater, its enforcement differs according to country. The European Commission has a policy of Television without Frontiers, but this is not yet worked out in detail. It is also a waste of money to advertise goods in areas where they are not on sale. While one technology to avoid this is available, the insertion of alternative material cannot be done effectively except by cable systems.

The question of what television standard to use has been hotly debated. The terrestrial systems (PAL, SECAM, NTSC) were developed from the older monochrome systems and suffer from certain defects due to the transmission of the colour and luminance information simultaneously within the same bandwidth (cross-colour and cross-luminance). The IBA (Independent Broadcasting Authority) in the UK developed a system called MAC (multiplex analogue components) which eliminated these defects and was claimed to be more suitable for transmission by satellite since these transmissions use FM, as opposed to AM, for terrestrial broadcasting. The MAC system eliminates the defects by transmitting the luminance and chrominance information at different times. The IBA argued that with the introduction of a new transmission medium the opportunity should be taken to adopt a better system, and that it would be better for all countries to adopt the same system to reduce manufacturing costs.

The MAC system was adopted by the UK Government for DBS in the UK and also adopted by the European Commission. Unfortunately, the first system to use MAC, British Satellite Broadcasting, failed commercially and was taken over by Sky which is wedded to the PAL system. The MAC system is in use in France and Scandinavia, and it includes the possibility of providing better pictures on the normal 16:9 format in part by use of digitally assisted techniques to improve resolution. Time will tell whether these techniques will come into use, or whether completely digital systems will be used for high definition TV (HDTV) (see section 52).

There are several variants of the MAC system depending on how the associated sound signal is transmitted. In the two principal systems D-MAC and D2-MAC, the sound is transmitted in digital form in what would be the line synchronizing interval of the vision signal. In D-MAC the digital information rate is 20.25 Mbit/s and in the D2-MAC system it is 10.125 Mbit/s. The D2-MAC system is more suitable for cable systems, and the D-MAC system, which was adopted by British Satellite

Broadcasting, is not now in use, except on the BSB satellites which will probably cease transmission at the end of 1992.

Some satellite systems depend for their financing on advertising, while others depend on subscriptions. It is clearly necessary to deny access to programmes to people who have not paid the subscription, and this has led to the development of a number of scrambling and encryption systems. The most effective of these cut each line into two parts and then interchange the parts (cut-and-rotate). The cut point is changed every line in accordance with a prearranged sequence, which makes it extremely difficult to "hack" the signal. The management of the encryption keys is complex and since the broadcaster's income depends on the security of these systems, they are seldom shared between broadcasters.

Part I
Strategic Issues

54 Telecommunications Network Numbering

Section 44 on direct dialling in to PABXs indicated the potential for problems in shortage of subscriber numbers in existing network numbering schemes. In section 48 on freephone and 800 series services we indicated an earlier UK solution to the problem which was to cease to include a letter significance of the exchange name in the number. At about the same time that the UK effected that change, the Americans introduced area codes as well as exchange identities; in America and Canada at the moment one dials 10 digits;

> ××× for the area
> ××× for the exchange and
> ×××× for the subscriber.

The exchange significance is notional as the typical size of an American urban exchange may exceed 50 000 lines and would therefore use several exchange codes.

In both the UK and the North American numbering plan, although the alphabetic significance of the numbers is not used, there is still a geographical significance to the number. 272 in England indicates numbers in Bristol city.

Network numbering is standardized, to an extent, by the CCITT.

Figure 54.1 shows a synopsis of the recommendations defining international numbering. CCITT Blue Book Recommendation I.330 (not shown in the diagram) describes ISDN numbering and addressing principles. Recommendation E.164 describes the numbering plan for the ISDN era and identifies the need for interworking arrangements between ISDN and present dedicated networks. Recommendation E.165 sets a specific time, 31 December 1996 at 23 h 59 m Coordinated Universal Time (UTC), after which all ISDNs and PSTNs can use the full capability of Recommendation E.164.

The ISDN basic access can accommodate up to 8 terminal devices. The ISDN concept includes the possibility of a user being identified at any point in the network. Existing developments such as direct dialling in (DDI) to PABXs and Centrex will be supported and encouraged by the

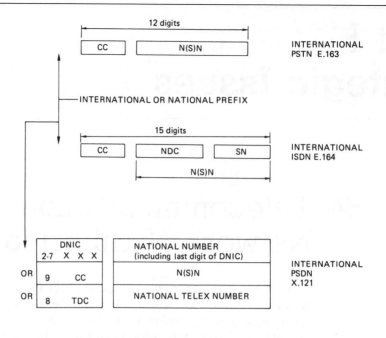

CC PSTN or ISDN Country Code
N(S)N National (Significant) Number (i.e. excluding prefixes)
NDC National Destination Code
SN Subscriber Number
DNIC Data Network Identification Code
or 3-digit data country code different from CC
TDC Telex Destination Code

Fig. 54.1
International network
numbering

ISDN. All these aspects will cause an explosive growth in the requirement for unique network numbers. At present in the UK, subscribers to Mercury must have a different directory number to that required for the BT network. Thus an organization with PABX exchange lines incoming from both BT and Mercury must have a different directory number for each group. Given adequate facilities in the SPC local exchange for number interpretation, there is no technical reason why this should be so.

All these considerations lead to a requirement that the public exchanges in the modern, liberalized, network incorporating ISDN must provide facilities for examination of all, or most, digits of the network number. The national and international numbering schemes must be altered as necessary to provide for an adequate supply of unique numbers. Just recently, the UK has implemented a plan for London area numbering which replaces the (0)1 prefix with (0)71 and (0)81.

National and world numbering schemes have become an important human resource and it is of concern to all of us that this resource is conserved. Not all recent decisions give confidence that the authorities are fully aware of their responsibilities in this regard. A very old telephone number, 01 368 (ENTerprise) 1234, the number of STC New Southgate, now Northern Telecom Europe (used as an example in the "bible" of UK telephone engineers, Atkinson's *Telephony*) has recently been changed to 081 945 4000. This is because STC has chosen to use Centrex on the alternative MCL network. There is no good technical reason why the same number should not be used for the same termination on either network but the allocation of a different exchange code, 945, to the MCL local exchange avoids the need to require existing BT exchanges to examine the subscriber number. This unnecessary proliferation of codes is, however, a misuse of the national numbering resource.

With this background the decisions being taken at the moment in both North America and the UK to extend the numbering schemes give cause for some concern.

In the UK, OFTEL commissioned a study and published the results. The most economical and sensible suggestion was a gradual loss of the geographical significance of the area code allied to a committment to extend the numbering scheme to 10 digits in the future in a gradual manner. The option chosen and now announced for implementation by 1994 has been to add an extra digit to the front of all numbers permitting the initial digit to indicate a specific network or service. This abandons any attempt to rationalize the present proliferation of unnecessary numbers for different networks and services and adheres to a geographical significance which is irrelevant in the presence of the personal portable numbering available with the ISDN.

The North American approach is perhaps more sensible, although there too there is not any attempt to combine numbering across networks and services. There the area code is being deprived of its geographical significance in a gradual manner and there is a plan for the introduction of a fourth digit to the area code.

55 Telecommunications Standards

International telecommunications, practically alone in international relations activities, has been conducted successfully and amicably since 1868 on a basis of unanimous consensus agreement. The International Telegraph Committee (CCIT), originated in that year, grew with the technology it controlled into the CCITT (telephones and telegraphs) and the CCIR (radio), and these in turn became part of the ITU when the communicators were assumed as agencies of the newly formed United Nations.

It has been the practice of the ITU for very many years to arrange its affairs into a 4-year cycle. During each period outstanding questions and new problems are considered by Study Groups and each cycle culminates in a Plenary Session at which the new and revised Recommendations agreed during the period are unanimously ratified. The current recommendations, the Blue Books, are the product of the ninth such plenary in Melbourne in 1988. They may also be the last complete series published in this cyclical fashion. One of the decisions taken at Melbourne was to permit immediate publication of any Recommendation which achieves consensus agreement in its parent Study Group. CCITT Recommendations are now appearing individually, presently using a white cover. At the next Plenary it will remain to recognise their publication and to agree by consensus any Recommendations which failed to achieve consensus within their Study Group.

The CCITT reached this decision because it was felt that the vastly increased body of work and the increased importance of the Recommendations demanded a quicker publication cycle. It was argued that other bodies using majority voting and public enquiry achieved faster delivery. This argument is however difficult to prove.

The subject matter of CCITT recommendations was confined, until quite recently, to areas pertinent to international communications only. Thus CCITT could define a budget of loss for an international telephone connection but could make no recommendation on how the portion of the budget destined for the national part of the connection could be apportioned. Similarly, CCITT R2 Signalling is defined for international use only, leaving considerable freedom for national variation. This deliberately confined stance of the CCITT, often criticized by engineers finding difficulty reconciling their systems to national variations, has, because of the modern need for defined protocols which operate from user to user, such as the ISDN protocols, been abandoned in favour of attempts to establish uniformity throughout the network and not only, or even principally, across frontiers.

This increased penetration of the CCITT remit was first evident in digital network definition (G.732, etc.), followed by common channel signalling, particularly CCS 7(Q.701, etc.) and by data communication (V. series and X. series) and now by the I. series on the ISDN.

The extension of the CCITT remit, while being caused by the increasing complexity of the protocols requiring standardization, has been also as a result of popular demand. A significant aspect of this demand has been the fragmentation of the network provider authorities and the need, in liberalized networks, to seek approval for the connection of terminal devices to the network.

Figure 55.1 provides an overview of the international bodies chiefly concerned with communications.

One of the driving forces towards liberalization has been the European Economic Community (EEC). Their requirements for open provision of goods and services throughout the Community met with resistance in the telecommunications sphere because in each country a terminal had to satisfy a different set of technical standards before it was allowed to be connected to the network. This meant that the local industry had a distinct advantage and telecommunications exports between member countries have been low to vanishing point as a result.

In 1987 the EEC stimulated the creation of the European Telecommunications Standards Institute (ETSI) charged to produce European Norms to set common requirements for the whole of the Community.

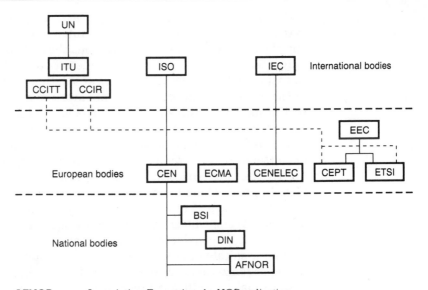

Fig. 55.1
International standards
bodies

AFNOR	Association Francaise de NORmalisation
BSI	British Standards Institution
CCIR	Comité Consultatif International Radio
CCITT	Comité Consultatif International Telégraphique et Télephonique
CEN	Comité Européene de Normalisation
CENELEC	Comité Européene de Normalisation de Electrotechnique
CEPT	Comité Européene de Postes et Telecommunications
DIN	Deutsche Industrie Norm
ECMA	European Computer Manufacturers Association
EEC	European Economic Community
ETSI	European Telecommunications Standards Institute
IEC	International Electrotechnical Commission
ISO	International Standards Organization
ITU	International Telecommunications Union
UN	United Nations

Some of these standards may then be used, where necessary, to form the basis of legally enforceable regulations.

ETSI operates in much the same way as CCITT through a number of committees and sub-committees analogous to the CCITT Study Groups. Like most other standards-making bodies, other than the ITU, ETSI relies on a voting procedure which, like everything in Europe, is weighted according to the economic "weight" of the member country. At earlier stages however consensus is required so that a draft standard must be agreed unanimously by its parent committee prior to being published for public comment and adopted, if necessary, by weighted vote of the members.

The discussion has described, and justified, the need for standards-making bodies to become far more involved in technology because technology's advanced complexity requires increased control. It has justified also the need for standards to increase the areas of free markets in the provision of telecommunications networks, services and terminals. Not all the results of this increased involvement have been universally happy however.

Because the network is to be universal and the complicated terminal is to be connected to the network at any point in the world, the terminal specification and network protocol must be compatible so that terminals may interoperate across the worldwide network. This is not a simple task to ensure and the present attempts at solution are biased towards design of the protocol by international committee *before* any trials of tentative realizations of possible protocols. The international bodies, hugely successful in standardizing telecommunications up to the present, are being asked to perform design tasks more proper to dedicated design teams uncluttered by considerations of national or organizational interest.

ETSI has perhaps provided the germs of a solution to this problem also. In areas where a distinct body of work is known to be required in order to produce a standard, ETSI recruits and hosts a team of experts, a Project Team, who are supervised by the parent committee and whose output will be considered for approval by that committee. This could well be a suitable pattern for future development projects where individual organizations are too unsure of the direction standardization will take to make significant investment on their own account. At present such ETSI Project Teams are confined to development of protocols for standardization, equivalent to paper design, although in areas such as Conformance Testing, Project Teams (not all of them ETSI) have developed the complete testing software. There is no good reason why, should the ETSI members agree, ETSI Project Teams could not be used for conceptual development projects funded by the members.

56 Telecommunications Network Management

Network management, as with all management, involves every activity associated with the telecommunications network from planning the network prior to its inception to ensuring prompt repair of a terminal equipment. The activities of the manager are also rather different between the management of the public network and the Telecommunication Manager in a business entity managing a private network. That having been said however, there are many Telecoms Managers in business whose networks are larger and more complex than many national or even regional telecommunications networks.

In the sections that follow we will deal with the operation and maintenance of the network and with the related problems and opportunities associated with the, generally better, reliability of modern networks. Of course, the two subjects are interrelated. We also deal with the public network problem of setting the correct tariffs for the services and the universal problem of forecasting future demand. In this introductory section we will look at network management overall, especially the absolute minimum that management of the network must involve.

The network, whether public or private, will have had to be designed to meet specific performance criteria, overall transmission loss across the network, traffic carrying capacity, etc. The Network Manager will monitor the network to ensure that these parameters continue to be met and will take timely action to upgrade the network as necessary to ensure its continuing performance. In designing the network, or more likely, in assessing the continued good design of an existing network, the manager will be looking for possibly rather obscure sources of potential trouble. Section 56B has a few cautionary tales to illustrate such dangers.

The private Network Manager, these days, probably has statutory obligations imposed by the regulating authorities to fulfil. Terminals must not be connected to the public network which may endanger the safety of other users or impair the performance of the public network. It could be, for example, that terminals that repeatedly attempt to re-try calls to busy users may not be allowed because they clog up the public network with unnecessary traffic. This is a problem, incidentally, that the ISDN removes or minimizes as the fact that the called user is free or busy is established over the common channel signalling path first, before any precious network connection resources are dedicated to the call. If the called user is busy then no connection is established although a message may be left at the called user's terminal.

Certainly the private network that carries public network traffic is required to provide transmission performance which is in line with the public network. The calling user, unaware that the call is passing over private networks, complains about lack of quality to the local telephone operator.

The Network Manager will be always looking at the network costs and the technical options for reducing these costs. The sections comparing leased line private networks and Virtual Private Networks illustrate some of the possible considerations.

We talk about a Network Manager when, in the case of the public network at least, we may really be talking about huge public or private corporations. The task gets bigger but the components of the task are similar to those we have described. Is the network performing to its optimum level; are the network nodes and equipments communicating satisfactorily over compatible signalling systems; are the network management information and control mechanisms available centrally and locally in the format that will allow quick and effective action to solve problems?

Network Managers of private networks have, increasingly, professional bodies to whom they can look for support and advice. In the UK the Telecommunications Managers' Association (TMA) performs this function as well as fulfilling a more political role in representing the business user's views to government and the public network providers. A similar role is met internationally by the International Telecommunications Users Group (INTUG).

The TMA is a member of ETSI and the TMA and INTUG participate in CCITT meetings as members of national delegations. There is a limit to the coverage they can give to these debates because of inadequate funding. There is a convincing argument for Telecommunications Managers of the larger businesses with extensive international networks to consider taking part in these international bodies on their own account. Too often at these debates the interests of the network provider and the manufacturer are well represented but little consideration is given specifically to the needs of the user.

56A Centralized Maintenance Systems

Most modern telephone exchanges require no maintenance adjustments: there are in general no moving parts so there is nothing to lubricate, nothing to clean, polish or readjust. When networks change over from electromechanical space-division switching and analog transmission systems to computer-controlled solid-state time-division switching and digital transmission systems, all the administrative, operational and maintenance functions can be streamlined.

Operational and maintenance services normally comprise four basic functions:

1) Service management, i.e. control of service offered to subscribers.
2) Maintenance management, i.e. the maintenance and repair of exchanges, external plant, transmission equipment, subscribers' apparatus. Software may also need attention.
3) Network management, i.e. management and supervision of traffic flow; short-term and long-term planning.
4) Call accounting, i.e. control of all aspects of recording information in relation to the bills to be sent to subscribers.

Because there is nothing to adjust and the number of faults is likely to be very small, it is not usually necessary for modern exchanges to be manned by permanently stationed maintenance staff. Most national administrations now therefore centralize their maintenance activities, merely sending technicians out to visit exchanges to change faulty items. Most types of modern exchange incorporate diagnostic programmes which not only localize their own faults (and give a print-out, describing the card or unit which is faulty and should be changed) but rearrange the traffic handling pattern of the exchange so that any such faulty section is by-passed. There is usually sufficient redundancy built into the exchange design to maintain traffic handling capacity during all except the most severe faults; it is indeed probable that most subscribers on an exchange will not even know that a fault has developed, been localized and cleared unless they happen to see the technician's van parked outside the exchange building.

This centralization of maintenance efforts means that in the unlikely event of an exchange developing particularly severe problems, it is possible to bring greater resources to that exchange to ensure that normal service is resumed with the minimum of delay. The organization of the maintenance effort into a single large group also allows expertise to be shared and ensures that individuals are continuously exercising and improving their skills.

Most administrations complement their centralized maintenance operations by centralizing all repair activities also. It is rare for maintenance staff at a modern exchange to be required to test and repair faulty printed circuit boards. First line maintenance normally takes the form of substitution of a new printed circuit board using information provided by the exchange itself and sent as a data message to the remote operations and maintenance centre. The replaced unit will then normally be sent to a specially equipped repair centre for repair. Here it must be borne in mind that many printed circuit boards now incorporate microprocessors so if a board is suspected of being faulty it has to be tested by a sophisticated and expensive computer-controlled test set. This test set will have been programmed to test all the functions of that particular board and of any other boards which might be sent to the repair centre for test and repair.

By the time any exchange has been brought into service, all its software should have been exhaustively tested, "de-bugged", and proved satisfactory. A master copy of each exchange software is usually held available for reference when necessary, in secure and non-volatile stores. It is common practice for wanted information to be called forward from one store to another, for example into an immediate access store. If software is suspected of being faulty (one plane of action may give a different answer from others), then the software concerned is usually replaced by a new version of the same software (in case the original has become corrupted), by calling forward from a "clean copy" held in a secure store. This procedure is called software reload or rollback. It is usual for there to be several levels of this, depending on the nature of the difficulty encountered—it would clearly be unnecessary for the whole of the exchange's software to be changed if changing only a single small module would eliminate the trouble. These software reload actions are normally carried out completely automatically at exchanges themselves, not on a centralized basis. Only if faults occur which cannot be cleared by changing over to a locally-available clean copy of the relevant software is reference made to a centralized maintenance organization. In real emergencies it is however often possible to connect an exchange not merely to a regional maintenance centre but to a national research laboratory or to the manufacturer's design engineers, so that they can investigate the problems from the remote centre.

56B Network and System Reliability

The last section introduced the much increased reliability of modern systems. It is perhaps fortunate that the systems also introduce the capability for remote operation and maintenance because this has become essential in order to keep the maintenance force in training for their task. When failures occur so infrequently and, when they do occur, can often be solved by the simple replacement of a unit, there is very little happening to give the maintenance engineer sufficient experience of failure modes. Thus it is not only convenient but necessary to extend the maintenance area to cover sufficient equipment so that there is enough going wrong, even with modern reliable systems, for the engineer to learn from experience of actual failure.

Another aspect of this increased reliability is that the systems are realized very largely by means of software modules. The software itself will have been exhaustively tested during the manufacturing phase to exclude all but the most abstruse errors. There is however a danger that a route through the software process is so rarely taken that it has not been tested in the software emulation phases. When it does happen in practice the maintenance man is faced with a completely unexpected new occurrence in which it may be very difficult to trace the source of the trouble or even to cause the fault to re-occur in order to study the system behaviour.

The United States has been using a common channel signalling network interconnecting all its stored program control telephone exchanges for many years. The network is illustrated in fig. 56B.1 showing the considerable degree of back-up CCS links used to ensure that the network will be always available. Despite this provision for network security there have been several failures of the CCS network in the past few years, two of which were sufficiently serious to deprive much of the country of telecommunications for most of a working day.

These faults have occurred when a new release of CCS software had been installed in some or all of the exchanges and where a software fault caused one exchange to fail. Because this exchange broadcast the fact that it had failed as a *link* failure, without any further fault, surrounding exchanges concluded, wrongly, from this failure that they too were in error and announced their failure. The network therefore progressively collapsed.

Paradoxically, the situation would not have occurred in networks, such as the CCS network in the UK, which use a simpler back-up interconnecting pattern not so heavily interconnected in order to avoid error.

This is one of the cautionary tales, promised in the previous section, to underline the need for network managers to validate the networking and the nodes which is essential for their networks to remain operational. They must also be very careful to specify and supervise the introduction of new features and facilities which, if in error, may have a disastrous effect on their networks.

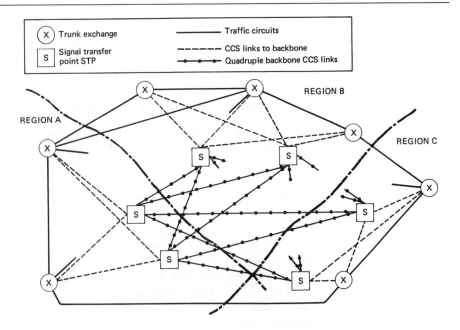

X	Trunk exchange	———————	Traffic circuits
S	Signal transfer point STP	- - - - - -	CCS links to backbone
		•-•-•-•	Quadruple backbone CCS links

REGION A

REGION B

REGION C

Fig. 56B.1
United States
quasi-associated CCS
architecture

The second tale is much more mundane, but just as important and, again, is an American experience. To save on the costs of energy it has become the practice in the US for operators to switch the exchange power supplies over from the mains to their own back-up generators. It is only in an economy such as America's, where oil fuel prices are subsidized, that such a method of saving could be contemplated. In late summer 1991 a main New York switching centre performed this automatic switch-over for the working day to the directions of a time clock setting. The maintenance force were away on a training course and nobody noticed that the generator did not start. A number of exchanges in the building, including transit exchanges carrying air traffic control traffic between the three main airports, therefore continued to operate for some 5 hours until their stand-by batteries discharged and much of New York and all air traffic came to a halt as a result.

Network management at its most basic can sometimes fail; the manager has to look both for the most abstruse and for the glaringly obvious. Reliable networks are a function, not just of reliable component systems, but of their systematic assembly into reliable networks, and of eternal vigilance on the part of the network manager, vigilance extending from the design of the network through network events to the whereabouts of the network technicians.

56C Telephone Tariffs

1 Viability Study and Tariff Structures

Capital costs (line plant, switching equipment, buildings and land, etc.), financing costs (interest on loans, etc.) and operating costs (staff salaries, maintenance and power, etc.) all have to be taken into account when calculating the total expenditure involved in providing telephone services. If the administration is to stand on its own feet financially, all these costs must be recovered; it is normally also necessary for a profit to be earned.

In some countries a financial target is set by government; it is therefore necessary for a tariff structure to be established so that the telephone system will earn specified profits. In other countries the tariff structure is fixed on purely political grounds by deciding how much the people can reasonably be called upon to pay for telephone service. Some countries may in fact subsidize telephone services by providing them for citizens at less than their true total cost, as a social measure.

A viability study is usually prepared in draft form at an initial planning stage and then refined when major tenders have come in and been evaluated so that actual expenditure during a main development scheme can be considered, instead of basing calculations purely on estimates and forecasts.

Various techniques can be employed in measuring viability; discounted cash flow and return on capital techniques are the principal measures. A study will usually end by recommending rates of return considered likely to be economically acceptable.

Comprehensive telephone tariffs have to be designed to generate sufficient revenue to enable the system to meet its required rate of financial return. The objective is usually one of maximizing the use of the system while meeting revenue targets.

The advantages and disadvantages of the various methods of charging for local telephone service have to be considered; these methods include:

a) Full flat rate—no separate charges for calls made.
b) Limited flat rate—some calls charged for (e.g. a free call allowance or all calls within a specified area free of charge).
c) Message rate, untimed—charge per call irrespective of duration.
d) Timed measured rate—charge per unit of call duration. This is sometimes called usage-sensitive charging.

Combinations of these different methods are possible, particularly in a large system.

The most suitable method for implementation then has to be selected and a network call charging plan produced which will conform with the technical capabilities of the system. Tariff proposals not only have to cover the charges for exchange line service and for calls, but must also include the charges for additional apparatus such as private branch exchanges and extension telephones. These charges normally have to be cost related so that no one service is "carried" by others.

2 Metering and Billing

Administrations which charge for local calls usually meter them by associating a simple cyclometer type meter with each exchange line. Meters are normally mounted about 1000 to a rack; they are read (sometimes photographically) and bills prepared on the basis of current reading less previous reading (less the number of test pulses used in the period). Newer exchanges can provide the same meter readings by storage of totals on a software basis; the cumulative totals can be read out whenever necessary.

There are three basic methods of using ordinary subscribers' meters for trunk and junction call charging:

1) Multimetering: when the called subscriber answers, the calling subscriber's meter is stepped up a specific number of times, depending on the exchange to which the call is routed. No timing is involved.

2) Repeat Multimetering: when the called subscriber answers and every 3 minutes thereafter, meters are stepped up by a fairly rapid train of pulses, e.g. six pulses (in 3 secs) (if the charge is 6 × local call fee). Pulses are generated within the exchange and operate straight to the meters. The particular pulse rate given depends on the exchange to which the call is routed.

3) Periodic Pulse Metering: PPM involves the selection of a pulse rate which is dependent on distance, e.g. calls from London to France could be pulsed once every 7.2 seconds, once every 4.8 seconds to the USA, once every 2.4 seconds to Nigeria, once every 20 seconds for a daytime local call during peak periods, or once per minute during cheap-rate periods, and so on. These timing intervals can rapidly be changed, making fine control of revenue possible.

All three of these systems involve bulk billing. The same meter is used for local calls as for trunk calls so there can be no separation of charges.

The ordinary electro-mechanical subscriber's meter itself will operate satisfactorily about 3 times per second, and metering in software could be many times faster than this, but fee determination is often carried out at an exchange other than that of the originating subscribers, so pulses have to be sent over junctions back to the originating exchange. Metering over junction equipment cannot be expected to operate without distortion more rapidly than 1 pulse per second.

Bulk billing has advantages of simplicity for the administration but there are disadvantages, both for the customer and the administration:

a) International Subscriber Trunk Dialling (ISTD) and national STD charges have to be closely related to local unit-call charges, so rates cannot be changed (e.g. to take account of currency exchange rate changes) unless pulsing equipment has suitable spare outputs readily available or can readily be changed or modified.

b) Limitation to a maximum of 1 pulse per second (i.e. a maximum charge of 60 local call fees per minute) is insufficient to cover costs in areas where the local call tariff is held down by Government action and where distances are large and there are only small numbers of circuits in traffic groups. In Europe and in the USA, circuit groups are big enough to give real economies of scale, whereas in some territories demand is relatively small and can only justify the use of comparatively expensive 6 or 12 channel systems.

c) The presentation of an itemized trunk call account is increasingly demanded by customers.

Bulk billing can produce a great many complaints by customers (and resultant non-revenue earning investigations by administrations), especially when expensive international calls can be dialled direct.

Several methods of automatically preparing itemized bills for self-dialled trunk calls have been developed. These can be divided broadly into systems in which some sorting is done at the origin and those in which sorting is done at a computer centre which can be remote from the exchange. There is some confusion about definitions but, in general, Automatic Message Accounting or AMA refers to the whole procedure, toll ticketing to a system with local sorting, and Centralized AMA or CAMA to a system with remote sorting:

a) Toll Ticketing: all the information necessary to prepare the bill for each call is punched on a single card or recorded in one stretch of magnetic tape (e.g. calling number, called number, time on, time off, etc.).

b) Centralized Automatic Message Accounting (CAMA): all required information about calls is recorded as it happens (there is usually a small temporary data store only), usually on magnetic tape which then has to be processed by a computer to link together "time on" for a particular call with "time off" for the same call so that chargeable time may be calculated.

Modern stored program control systems allied to common channel signalling make it possible, and relatively easy, to provide itemized bills. In view of the increasing demand from the public for telephone bills, like most of the other bills we receive, to detail how we have spent our money, itemized bills will become the norm in the future.

The preceding description of the "traditional" metering arrangements is still important however as it has fundamentally affected the way even the most modern systems are designed. For many years to come, call charging information will have to be transmitted across the network in the form of periodic pulses or their CCS equivalent in order to cater for the old-fashioned bulk billing equipment still existing in the network.

The description also serves the purpose of illustrating how a significant proportion of the complexity and cost of telecommunications systems is caused by the need to record the charge for the service. This is becoming even greater as we introduce the multiplicity of services possible with the ISDN. It is not altogether obvious that this increased complexity is cost effective. There could be good arguments for offering services as a free extra and collecting profit from the increased calls made because of the services. Such a review has not, as yet, been undertaken but could well revolutionize the attitude of administrations to the provision and charging of services.

56D Forecasting Future Demands

It was undoubtedly a telephone system planner who first said that forecasting was always difficult, particularly forecasting the future. Telephone people are especially sensitive to forecasting difficulties because someone always has to take the blame when telephone service cannot be provided in any given location immediately it is requested. Unfortunately even what seems to be a simple request for service sometimes behaves like the last straw and breaks the camel's back.

In most countries it takes between five and seven years to get a new telephone exchange into service, starting from the selection and purchase of the site (sometimes this means that a lengthy public enquiry has to be held), the design and construction of the building, the design, manufacture and installation of the switching equipment itself, the selection and training of all staff needed, the construction of all necessary manholes and duct routes so that cables can be fed into the new exchange, the planning, purchasing, laying and jointing of all these new cables. And many existing exchanges will be affected by any proposed new exchange and will themselves need extra equipment to be designed, purchased, installed and tested. So the time passes very rapidly.

Planning a new telephone system normally begins with the preparation of a set of large-scale maps of the area concerned showing every plot of land and every building. The area is divided on these maps into blocks about 100 metres by 100 metres in size. All available information is collected together about possible new road construction schemes, housing schemes, office blocks, long-term municipal development plans, and so on. Visits are then paid by skilled staff to each of the blocks. Starting with up-to-date information about the number of existing telephone lines, telex lines, etc., in each of the buildings in the block, plus the numbers on waiting lists for service, estimates are made of the probable numbers of lines likely to be needed in 5, 10, 15 and 20 years' time, based on a study of all available planning information. These forecast figures are duly entered in boxes on the master map.

Study of these figures will indicate the preferred location for an exchange on the basis of it being the "copper centre": the location which could serve all the forecast lines at minimum cable cost. The practical copper centres often move across a city with the passage of time, e.g. with the development of an industrial estate in the suburbs. Modern transmission and switching practices, such as the use of digital exchanges and PCM systems to serve remote concentrators located in outlying parts of a city, nowadays make the theoretical copper centre of less direct relevance when a new site has to be selected but, if there is a choice of sites, the different total costs of cabling needed for each possible exchange site all have to be considered before reaching a decision. Civil works, including the laying of duct routes, the construction of manholes, etc., are usually planned on the basis of 20 years' life; it becomes extremely expensive if roads have to be excavated and surfaces reinstated every few years. Primary distribution cables are then pulled in to these newly laid ducts; usually enough cable pairs are provided initially to meet the 5 year forecast of demand for the area concerned. Secondary distribution cables, the smaller cables feeding from flexibility cabinets out to distribution points, are often directly buried in the ground, i.e. without using ducts. In such cases an attempt is usually made to provide sufficient cable pairs to meet the demand forecast for at least 15 years ahead. If the information available to the original planners was incomplete, or if major changes of land use are agreed and locally implemented at short notice, it is no wonder that there are sometimes insufficient cable pairs available to meet current demands for service.

The expansion of complete national systems can best be looked at on

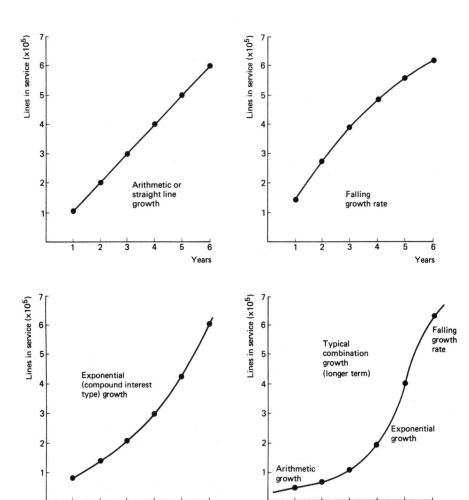

Fig. 56D.1
Types of demand curve

an overall basis. If demand for service is plotted against time using ordinary arithmetic scales, it is usual for this curve to take one of a number of shapes, depending on many economic factors and on the length of the period being studied (see fig. 56D.1):

 a) Straight-line growth rate.
 b) Exponential or compound interest type growth rate.
 c) Falling-off in growth rate.
 d) A combination of all three of these.

Most developing countries experience an exponential or compound interest type of growth of demand; it will be seen that it is extremely difficult to use a graph of this nature to extrapolate forward to give forecast figures for later years, since the line becomes almost vertical (fig. 56D.2). Here the usefulness of semi-logarithmic graph paper becomes apparent: when compound-interest type growth is plotted on this paper a straight line results which can legitimately be extended forward to provide forecast figures for future demand (fig. 56D.3).

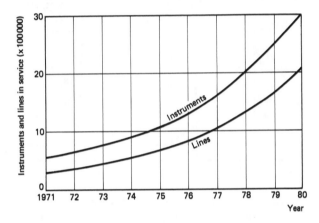

Fig. 56D.2
Typical figures for the growth of an actual national telephone system: growth plotted on linear scales

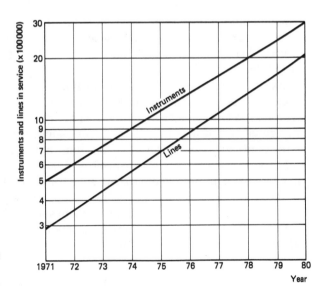

Fig. 56D.3
As fig. 56D.2 but growth plotted on semi-logarithmic paper

Planners often have to decide which way to tackle a job in order to obtain the best value for money. One simple example is given in fig. 56D.4—here the choice shown was between

1) Installing new analog exchanges.
2) Installing new digital exchanges.
3) Installing new digital exchanges using remote concentrators.

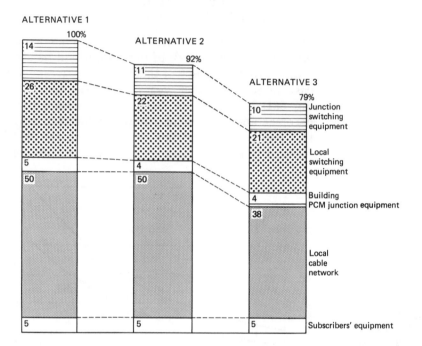

Fig. 56D.4
Cost comparisons of various network planning alternatives

The third alternative clearly provided all the necessary service features at a capital cost of only 79% of the original "traditional" analog scheme.

In an ideal world, good planning would be followed by the installation of just enough equipment to satisfy all demands for service, as they occur. In our real and imperfect world there are however usually a few hiccups. Actual development on the ground all too often requires different telecommunication services (or more of them) than was envisaged by the planners. Also it is not often possible to provide telecommunications equipment in small packets, keeping just ahead of demand. For example, if a trench is to be excavated along a road and ducts and manholes constructed it is usually economic—and certainly much preferred by residents—for such works to be done not more often than once in twenty years or so. This necessarily means that many of the ducts laid will be empty and not earning their keep for several years. The planner's aim must always be to keep this "burden of spare plant" as low as possible and to use technological improvements and advancements not only to improve services provided for customers but also to reduce the capital and running costs of the system.

57 Telecoms Security

In the early days, computers were rather like temple-gods: input information, usually offerings in the form of piles of punched cards, had to be physically brought in (often on slippered feet) to the Presence, usually in a closely guarded air-conditioned room.

There were no real security problems in those days; locking the door and ensuring that power supplies were smooth and uninterrupted were sufficient. As soon as it became possible to use telecommunications links so that a computer could be remotely controlled, or terminal users on the computer in one office could access databases via another computer, all sorts of problems began to arise.

It is relevant that companies everywhere are happy to use modems and relatively cheap dial-up circuits on the ordinary public switched telephone network for many of their inter-computer links. The alternative is leasing expensive links on a full-time basis—but sometimes a link is needed for only a few minutes use, and not every day at that, so leases are very expensive options. Telephone authorities all over the world do their best to encourage customers to make maximum use of all available services (even chat-lines of doubtful social value) so this explosive growth in data traffic carried over switched networks to which the general public has unrestricted access is seen as normal and acceptable.

Even if your office computer itself has no dial-up public-access outside lines at all, and works all its inter-computer links on permanently-leased private data circuits, the probability is high that somewhere in the associated network there will be a gateway to a public system, either a packet switched data system or the ordinary circuit-switched telephone system. Finding these entry points is not always at all difficult for a keen hacker, an electronic trespasser who accesses other people's computer networks for unfriendly purposes. Hacking itself is not illegal in some countries, but the results of hacking are sometimes outside all countries' laws.

Most hackers start off as enthusiastic computer-literate students, using a personal computer, a modem, and a telephone line. They are very proud of themselves when they work out exactly what series of actions they have to take to put their PC directly in touch with a big company's main computer. Most companies ask for passwords to be given before they will let you get any further, but keen hackers found, all too often, that the most obvious of passwords were often adequate—the Managing Director's initials or the Chief Engineer's date of birth were commonly all that was needed—and few companies change their passwords often enough to introduce real difficulties. It is even possible in some computers for very competent hackers to get straight in to the Network Manager's file where lists of users' names and their passwords are stored—these lists can then be sold to other hackers.

In the beginning it was all a bit of a joke; if you managed to get through the first hurdle you sometimes keyed in as a souvenir of your uninvited visit a short cryptic message of the "Bristow was here" type, then cleared down and readied yourself for another challenge to your intellectual ability.

As companies began to use their computers more and more for financial and project planning work (and not solely as a pay-roll producer) it became clear that there was a market in the rather murky world of commercial espionage for all sorts of computer-stored information, and many hackers were by now well placed to supply this market with all that it needed.

Some hackers are however not content to be merely trespassers and eavesdroppers, they leave terrible souvenirs behind after they have attacked and penetrated an unsuspecting computer. The forms of attack on computers which are most often encountered are logic bombs (or trap-doors), Trojan Horses (or software moles), and computer viruses.

● *Logic bombs (trap-doors)* These are small sectors of a computer program that can be triggered off (or swung open) by a word or sequence of characters that the hacker has inserted into the computer's software. In the 1970s the US Navy reportedly found that a trap-door had been inserted in the computer of a warship: immediately a plane picked up by its radar had been interrogated and identified as being an enemy plane the trap-door was opened and spewed out a planted subversive message giving false guidance information to the ship's missiles, ensuring that they would blow up long before reaching their target. A civilian equivalent example would be a logic bomb producing a modified copy of a company spreadsheet giving deliberately incorrect figures; a "classic" bomb would bring all the figures in some columns to zero, but only at specific times, just to make the bomb confusing and difficult to trace.

● *Trojan Horses (software moles)* These are false commands inserted inside legitimate commands, i.e. programs that perform services beyond their specifications. A naval Trojan Horse actually reported was one which instructed a ship's computerized command system to ignore any anti-submarine warfare commands given in a real combat situation at certain latitude and longitude coordinates. During exercises with dummy ammunition in the sensitive area there was no interference with any activities but, as soon as real action was needed and live ammunition was called for, all the commands were completely ignored by the command computer. A civilian example of a Trojan Horse is a program which purports to provide data compression but in fact deletes complete files of information from memories.

● *Computer viruses* These are programs which are able to copy themselves to other files, or disks, or to other computers, without specifically being called upon, so they are usually modifications of some commonly used program such as the operating system itself. Cheap pirate software disks have been one of the common sources of virus infection. For example, one virus prepared in Pakistan several years ago has managed to spread all over Europe and America by installing a small part of itself in the bootstrap sector of disks while storing most of its commands in sectors of the disk which are electronically marked as bad, so that they will not be over-written by other programs. Viruses can have many different functions; those most commonly encountered are the random erasure or corruption of valuable files.

Viruses, moles, hackers and bombs are however not the only security hazards. Even in modern telephone systems you sometimes get crossed lines and hear other people conversing: this is usually due to a temporary fault, most often to an external cable fault, but clever villains can and sometimes do tap across telephone lines to overhear all that is going on. So if commercially valuable data messages are being transmitted along a circuit it is always possible that some unauthorized person may be able to eavesdrop on these messages, and may even be able to chip in, pretending to be the distant end answering. Whenever money or big deals are involved it is therefore desirable to encode line signals in some way so that the casual (or not so casual) eavesdropper will not be able to profit.

A few years ago very few people realized how widespread computer-related crime had become. The name itself is of course a bit odd: it perhaps shows that telecoms authorities have better PR teams than computer companies, since almost all these new crimes have only been made possible by the existence of telecoms services. When computers were isolated in their ivory towers it was not so easy to work the kind of criminal magic that nowadays gets blamed on our friends in the computer world. I think it would be fair to say that telecoms people merely provide burglars with free ladders up to every window—it is up to the computer people to fit good locks! Be that as it may, if a company loses a few hundred thousand pounds to an insider who exploits a weakness in computer security the company usually dismisses the villain but keeps quiet about its losses, which may in some instances have passed unnoticed for many weeks. A big company could perhaps afford to take this somewhat leniently negative approach. But with the increased use of EFTPOS (electronic funds transfer at point of sale) the ordinary person in the street can be directly affected by a shortfall in security. He or she is going to be very upset if the monthly bank statement shows the account to have been debited for the cost of goods not actually bought, and is going to demand immediate corrective action.

More and more use is therefore now being made of complex encoding systems. Encryption is the name given to the procedure which converts text or voice information into undecipherable form. When properly used both the source and the contents of each message are protected.

To understand one of the basic features of simple encryption assume first that all voice signals have been turned into a digital form using PCM (see section 13). Letters and figures in text messages are also turned into digital form. ASCII (American Standard Code for Information Interchange) is the most common of these codes in today's digital world; it uses a 7-bit code for each letter or figure. So all the different types of information flowing along a circuit can be considered to be in the form of a stream of bits, a stream of 1s and 0s. By a completely arbitrary decision this stream of bits (although representing voice or text in the original) can be considered to be broken up into large chunks, each of which then represents a large multi-bit number. Each of these large numbers can then be fed through a computer, multiplied by a secret number, then divided by another secret number, and only then are the resultant numbers, a new series of 1s and 0s, sent out to line, with no resemblance at all to the original stream of 1s and 0s. At the other end the reverse process is carried out: the received 1s and 0s are multiplied and then divided by these secret numbers, and only the final result is fed out to the demodulators and decoders which turn the digital line signals into the signals which the output is geared up to receive. There are of course many additional complications in practice but the basic principle of there being lists of Top Secret "key numbers" to be used on particular days is still usually followed. It was the regular supply to the Russians of US Navy key lists that kept the members of the Walker spy-ring in funds for two decades, and enabled the Russians to decipher most of the US Navy's top secret messages during this whole period.

With villains and spies all trying to muscle in on all the many streams of data traffic, there are bound to be some slip-ups but in general what one person puts wrong another can usually (in the end, regrettably sometimes not until after the horses have all bolted!) put right, so viruses and bombs and hackers can with due care be beaten: to continue the medical analogy, "Personal Computer Hygiene Rules" is now the name of the game.

Staff with access to computer terminals often plug disks in and play computer games during lunch breaks, but experience has shown that many cheap computer games disks carry viruses of one sort or another so it will often pay a big company to provide a stand-alone microcomputer into which the staff can plug any games they wish, from any source; the terminals on the main computer are switched off completely during break periods. It doesn't matter much if the single stand-alone micro does get infected with a virus; it is isolated and can in due course be cured.

As telephone systems everywhere are modernized, new common channel signalling systems are being brought into service. These are able to forward the calling subscriber's number to the called end, so your computer will be able to check this against its built-in list of authorized users and terminate the call if the caller is not on the list, or even hold the caller on the line so that a special investigation may be initiated, perhaps leading to the catching of a really big fish.

58 Telecommuting

In 1980 one of the authors was involved in the development of the ITT (now Alcatel) System 1240 digital stored program telephone exchange system. One of the development projects was to design the operator's sub-system. Much work went into developing the design requirements and one of these was seen as a need to permit individual operators to work from home or, at least, from a local office within a short bus ride from home. Thus, instead of the large, centralized, operator rooms with many operators sitting side by side at consoles, what was envisaged was individual operators still working as a member of a large team, but doing it from isolated consoles in their homes or in several small offices throughout the area or country.

While this was perfectly feasible in 1980 when this work was taking place and the design specifications were developed to a fairly mature level, the actual development never took place because of other priorities within the organization. To our knowledge, no such distributed operator system is in use although the telephone system in Canada does use a number of distributed operator offices.

The thinking behind the distributed operator proposal was the need for young mothers to find work which enabled them to keep an eye on the children. Some sacrifices were envisaged in that it would be harder to feel part of a team with no direct contact and little chance to chat with colleagues.

Another early and more successful attempt at "telecommuting" was introduced by the Xerox Corporation in the UK in about 1982. They were going through hard times and wished to make some of their senior people redundant while still keeping their services on a contract basis. Volunteers were called for to become "home workers". Each homeworker was equipped with PC, modem, facsimile, etc. and could work at home on an occasional basis for Xerox. The experiment proved quite successful and the use of homeworkers is still continuing.

There have always been tasks that have more conveniently been performed from home; insurance agencies are one such and area management for distribution organizations is another. Increasingly such people are equipped with PCs, modems, and probably connection to the packet switched network so that they can work at home but have all the facilities of their headquarters computer systems and databases at their disposal.

There are difficulties in working from home: the loss of the social environment of the work place and the need to measure the work done to ensure fair remuneration. The second problem is largely solved by the very computer-based packages that are needed for homework, project planning and scheduling packages, for example. The former either becomes less important as the difficulty and expense of getting to work increases or can also be substituted by other, quite pleasant means.

The weekly or other regular gathering of the team can become a very pleasant occasion.

These difficulties were covered with some pessimism in the third edition but the outlook is now more hopeful. It suggested the use of satellite offices. These now exist although their use is more for the peripatetic executive than for a small local team from a big company.

Clearly what problems there are in working from home are none of them technical. The technical means to do so are in place now and will only improve with ISDN, BISDN, personal communications and the other developments which have been described in this book.

Index